MODERN CHEMISTRY

SECTION REVIEWS

LACIE DUFFY

HOLT, RINEHART AND WINSTON

A Harcourt Classroom Education Company

Austin · New York · Orlando · Atlanta · San Francisco · Boston · Dallas · Toronto · London

Section Reviews Pupil's Edition:

Copyright © by Holt, Rinehart and Winston

Cover: Portion of the periodic table superimposed on a photomicrograph of crystals of the amino-acid dimer cystine. Photo by © Mel Pollinger/Fran Heyl Associates, NYC.

Printed in the United States of America

ISBN 0-03-057354-8

5 6 7   862   05 04 03 02

TABLE OF CONTENTS

CHAPTER 1 REVIEW
Matter and Change

SECTION 1-1

SHORT ANSWER Answer the following questions in the space provided.

1. _a_ Technological development of a chemical product often ____.

 (a) lags behind basic research on the same substance
 (b) involves accidental discoveries
 (c) is designed to understand a practical problem
 (d) is done for the sake of learning something new

2. _d_ The primary motivation behind basic research is to ____.

 (a) develop new products
 (b) make money
 (c) understand an environmental problem
 (d) gain knowledge

3. _a_ Applied research is designed to ____.

 (a) solve a particular problem
 (b) develop new products
 (c) gain knowledge
 (d) learn for the sake of learning

4. _b_ Chemistry is ____.

 (a) a biological science
 (b) a physical science
 (c) concerned mostly with living things
 (d) the study of electricity

5. Define the six major branches of chemistry.

organic- study of carbon compounds
inorganic- study of non-carbon compounds
physical-study of properties and changes of matter
related to energy
analytical-identification of components and composition of materials
biochemistry-study of substances and processes in living things
theoretical-the use of math and computers to understand
and predict chemical phenomena

SECTION 1-1 continued

6. For each of the following types of chemical investigations, determine whether the investigation is basic research, applied research, or technological development.

basic research **a.** A laboratory in a major university surveys all the reactions involving bromine.

applied research **b.** A pharmaceutical company explores a disease in order to produce a better medicine.

applied research
~~basic research~~ **c.** A scientist investigates the cause of the ozone hole.

technological development **d.** A pharmaceutical company discovers a more efficient method of producing a drug.

technological development
~~applied research~~ **e.** A chemical company develops a new biodegradable plastic.

applied research **f.** A laboratory explores the use of ozone to inactivate bacteria in a drinking-water system.

7. Give examples of two different instruments routinely used in chemistry.

• microscope

• triple beam balance

8. What are microstructures?

structures that cannot be seen with the human eye

9. What is a chemical?

any substance that has a definite composition

10. What is chemistry?

the study of the composition structure, and properties of matter and the changes it undergoes

CHAPTER 1 REVIEW
Matter and Change

SECTION 1-2

SHORT ANSWER Answer the following questions in the space provided.

1. Classify each of the following as a *homogeneous* or *heterogeneous* substance:

heterogeneous **a.** iron ore

homogeneous **b.** quartz

~~homogeneous~~ heterogeneous **c.** granite

heterogeneous (homo) **d.** soft drink

heterogeneous **e.** milk

homogeneous **f.** salt

homogeneous **g.** water

homogeneous **h.** nitrogen

2. Classify each of the following as a *physical* or *chemical* change:

physical **a.** ice melting

chemical **b.** paper burning

chemical **c.** metal rusting

physical **d.** gas under pressure

physical **e.** liquid evaporating

chemical **f.** food digesting

3. Compare a physical change with a chemical change.

physical change — ~~a characteristic that doesn't make~~ a change in a substance that does not involve a change in the identity of the substance
chemical change — a change in which one or more substances are converted into different substances

Name _____ Date _____ Class _____

SECTION 1-2 continued

4. Compare and contrast each of the following terms:

a. *mass* and *matter*

matter - something that has mass and takes up space

mass - amount of matter something has

b. *atom* and *compound*

atom - smallest unit of an element

compound - pure substance made from chemically bonded atoms of two elements

c. *physical property* and *chemical property*

physical property - a characteristic that can be observed or measured w/o changing the identity of the substance. Chemical property - the ability of a substance to undergo a change that transforms it into a different substance

d. *homogeneous mixture* and *heterogeneous mixture*

homogeneous - uniform composition and properties throughout

heterogeneous - different compositio and properties throughout

5. Draw a diagram that compares the arrangement of atoms in the solid, liquid, and gas state.

Solid   Liquid   Gas

6. How is energy involved in chemical and physical changes?

heat and light can cause substances to change

MODERN CHEMISTRY

CHAPTER 1 REVIEW
Matter and Change

SECTION 1-3

SHORT ANSWER   Answer the following questions in the space provided.

1. A horizontal row of elements in the periodic table is called a(n) _period_____.

2. The symbol for the element in Period 2, Group 13, is _Boron_____.

3. Elements that are good conductors of heat and electricity are _metals_____.

4. Elements that are poor conductors of heat and electricity are _nonmetals_____.

5. A vertical row of elements in the periodic table is called a(n) _group_____.

6. The ability of a substance to be hammered or rolled into thin sheets is called
   _malleable_____.

7. Would an element that is soft and able to be cut with a knife likely be a metal or a
   nonmetal? _non-metal_____.

8. Group 18 elements, which are generally unreactive, are called _noble gases_____.

9. At room temperature, most metals are _solids_____.

10. Name three characteristics of most nonmetals.
    gases @ room temp., poor conductor of heat & electricity,
    _brittle_____

11. Name three characteristics of metals.
    _lustrious, good conductors of heat & electricity, ductile_____

12. Name three characteristics of most metalloids.
    _solid @ room temp., lustrious, semi-conductor_____

13. Name two characteristics of noble gases.
    _unreactive, gases @ room temp._____

SECTION 1-3 continued

14. What do elements of the same group in the periodic table have in common?

similar chemical properties

15. What do elements of the same period in the periodic table have in common?

similar physical properties

16. You are trying to manufacture a new material, but you would like to replace one of the elements in your new substance with another element that has similar chemical properties. How would you use the periodic table to choose a likely substitute?

look for a closely related element in either a group or period.

17. What is the difference between a family of elements and elements in the same period?

families have similar chemical properties, periods have same physical properties

18. Complete the table below by filling in the spaces with correct names or symbols.

Name of element	Symbol of element
Aluminum	Al
Calcium	Ca
Manganese	Mn
Nickel	Ni
Phosphorus	P
Cobalt	Co
Silicon	Si
Hydrogen	H

MODERN CHEMISTRY

CHAPTER 1 REVIEW
Matter and Change

MIXED REVIEW

SHORT ANSWER Answer the following questions in the space provided.

1. Classify each of the following as a *homogeneous* or *heterogeneous* substance:

homogeneous **a.** sugar

homogeneous **b.** iron filings

heterogeneous **c.** milk

heterogeneous **d.** plastic

homogenous **e.** cement

2. Select the most appropriate branch of chemistry from the following choices to best describe each of the investigations: organic chemistry, analytical chemistry, biochemistry, theoretical chemistry.

analytical **a.** A forensic scientist uses chemistry to find information at the scene of a crime.

theoretical **b.** A scientist uses a computer model to see how an enzyme will function.

biochemistry **c.** A professor explores the reactions that take place in a human liver.

physical organic **d.** An oil company scientist tries to design a better gasoline.

analytical **e.** An anthropologist tries to find out the nature of a substance in a mummy's wrap.

analytical bio **f.** A pharmaceutical company examines the protein on the coating of a virus.

3. For each of the following types of chemical investigations, determine whether the investigation is basic research, applied research, or technological development.

basic research **a.** A university plans to map all the genes on human chromosomes.

applied research **b.** A research team intends to find out why a lake remains polluted.

techno develop applied research **c.** A science teacher looks for a paint that will allow graffiti to be easily removed.

basic applied research **d.** A cancer research institute explores the chemistry of the cell.

basic research **e.** A professor explores the toxic compounds in marine animals.

4. Use the periodic table to identify the name, group number, and period number of the following elements:

Chlorine; 17; 3 **a.** Cl

magnesium; 2; 3 **b.** Mg

Tungsten; 6; 6 **c.** W

Iron; 8; 4 **d.** Fe

Tin; 14; 5 **e.** Sn

5. What is the difference between extensive and intensive properties?

an extensive property depends on the amount of matter present, an intensive property doesnot.

6. Consider the burning of gasoline and the evaporation of gasoline. Which process represents a chemical change and which represents a physical change? Give a reason for your answer.

burning the gasoline is chemical because it gives a reaction (light, explosion) and the evaporation does not.

7. Describe the difference between a heterogeneous mixture and a homogeneous mixture, and give an example of each.

a homogeneous mixture's composition is uniform: sugar in water a heterogeneous mixture's composition is not uniform: granite

8. Construct a concept map that includes the following terms: atom, element, compound, pure substance, mixture, homogeneous, and heterogeneous.

CHAPTER 2 REVIEW
Measurements and Calculations

SECTION 2-1

SHORT ANSWER Answer the following questions in the space provided.

1. Determine whether the following are examples of observations and data, a theory, a hypothesis, a control, or a model.

   _____observ/data_____ **a.** A research team records the rainfall in inches per day in a prescribed area of the rain forest. The square footage of vegetation and relative plant density per square foot are also measured.

   _____observ/data_____ **b.** The intensity of the precipitation, the duration, and the time of day are noted for each precipitation episode. The types of vegetation in the area are logged and classified.

   _____control_____ **c.** The information gathered is compared with the data on the average precipitation and the plant population collected over the last 10 years.

   _____hypothess_____ **d.** The information gathered by the research team indicates that rainfall has decreased significantly. They propose that deforestation is the primary cause of this phenomenon.

2. "When 10.0 g of white, crystalline sugar is dissolved into 100. mL of water, the system is observed to freeze at −0.54°C, not 0.0°C. The system is denser than pure water." Identify which parts of these statements represent quantitative information and which parts represent qualitative information.

   _____quantitative -_____

   _____qualitative - white, crystalline, denser than pure wats_____

3. Compare and contrast a model with a theory.

   _____theory - broad generalization_____

   _____model - specific types of theories_____

Name _____ Date _____ Class _____

4. Evaluate the following models. Describe how the models differ from the objects they represent.

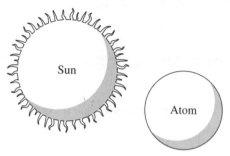

__model of sun-circular, fiery surface__
__atom- circular__

_____

_____

_____

_____

_____

5. __C__ There are _____ variables represented in the two graphs shown below.

    a. one   **b.** two   **c.** three   **d.** four

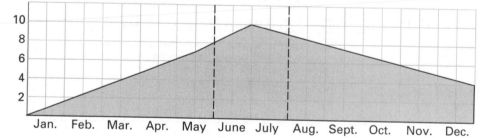

Name ______________________ Date ______________ Class ______________

CHAPTER 2 REVIEW

Measurements and Calculations

SECTION 2-2

SHORT ANSWER **Answer the following questions in the space provided.**

1. Complete the following conversions:

a. 100 mL = 0.1 L

b. 0.25 g = 25 cg

c. 400 cm^3 = 0.4 L

d. 400 cm^3 = 0.0004 m^3

2. a. Identify the quantity measured by each measuring device shown below.

a measures weight b. measures mass c. measures volume

b. For each quantity identified in part a, explain when it would remain constant and when it would vary.

inertia

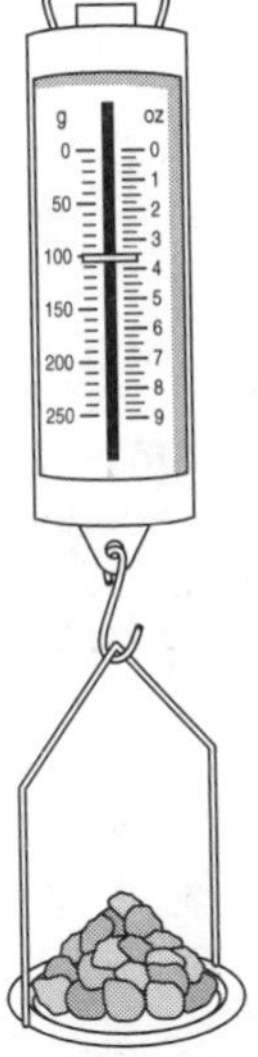

a.

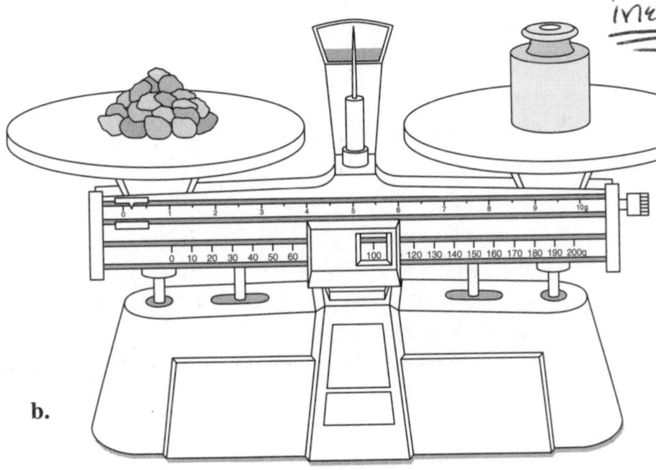

b.

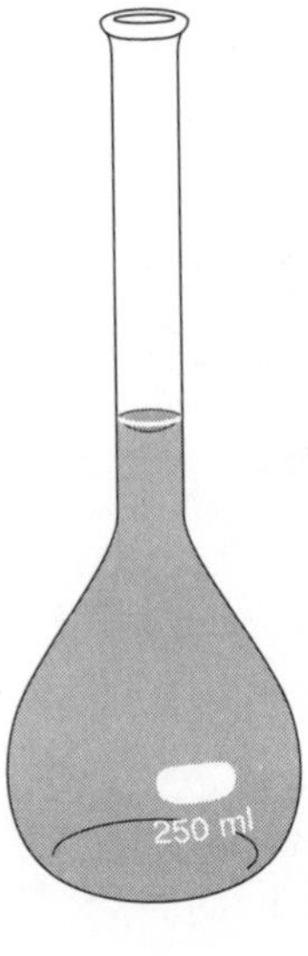

c.

Name _____ Date _____ Class _____

3. Use the data found in Table 2-4 on page 38 of the text to answer the following questions:

_____ sink _____ **a.** If ice were denser than liquid water at 0°C, would it float or sink in water?

_____ kerosene _____ **b.** Water and kerosene do not readily dissolve in one another. If the two are mixed, they quickly separate into layers. Which liquid floats on top?

_____ gasoline, ethyl alcohol _____ **c.** What other liquids would float on top of water?
_____ turpentine _____

4. Use the graph of the density of aluminum below to determine the approximate mass of aluminum samples with the following volumes.

_23.0g_ **a.** 8.0 mL

_40_
_~~4.05~~g_ **b.** 1.50 mL

_20g_ **c.** 7.25 mL

_9g_ **d.** 3.50 mL

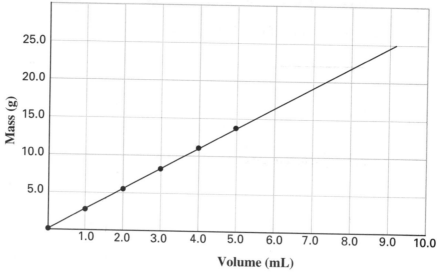

Mass vs. Volume of Aluminum

Mass (g) vs. Volume (mL)

PROBLEMS Write the answer on the line to the left. Show all your work in the space provided.

5. _27.0 g_ Aluminum has a density of 2.70 g/cm³. What would be the mass of a sample whose volume is 10.0 cm³?

$$d = \frac{m}{V} \qquad m = dV$$

6. _14 cm_ A certain piece of copper wire is determined to have a mass of 2.00 g per meter. How many centimeters of the wire would be needed to provide 0.28 g of copper?

CHAPTER 2 REVIEW
Measurements and Calculations

SECTION 2-3

SHORT ANSWER Answer the following questions in the space provided.

1. Report the number of significant figures in each of the following values:

___3___ **a.** 0.002 37 g   ___2___ **d.** 64 mL

___4___ **b.** 0.002 037 g   ___2___ **e.** 1.3×10^2 cm

___3___ **c.** 350. J   ___3___ **f.** 1.30×10^2 cm

2. Write the value of the following operations using scientific notation:

___10^{-1}___ **a.** $\dfrac{10^3 \times 10^{-6}}{10^{-2}}$

___4×10^{-2}___ **b.** $\dfrac{8 \times 10^3}{2 \times 10^5}$

___4.3×10^{12}___ **c.** $3 \times 10^3 + 4.0 \times 10^4$

3. The following data are given for two variables, A and B:

A	B
18	2
9	4
6	6
3	12

___inversely___   **a.** Are A and B directly or inversely proportional?

b. In the graph provided, sketch a plot of data.

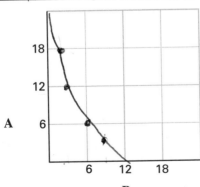

___no___   **c.** Do the data points form a straight line?

___$a \times b = k$___   **d.** Which equation fits the relationship shown by the data? $A \div B = k$ (a constant) or $A \times B = k$ (a constant)

___36___   **e.** What is the value of k?

SECTION 2-3 continued

4. Carry out the following calculations. Express each answer to the correct number of significant figures.

_____40.0_____ **a.** 37.26 m + 2.7 m + 0.0015 m =

_____2000_____ **b.** 256.3 mL + 2 L + 137 mL =

_____151_____ **c.** $\dfrac{300.\ kPa \times 274.57\ mL}{547\ kPa}$ =

_____100_____ **d.** $\dfrac{346\ mL \times 200\ K}{546.4\ K}$ =

5. Round the following measurements to 3 significant figures:

_____22.8_____ **a.** 22.77 g

_____14.6_____ **b.** 14.62 m

_____9.31_____ **c.** 9.3052 L

_____87.6_____ **d.** 87.55 cm

_____30.2_____ **e.** 30.25 g

PROBLEMS   Write the answer on the line to the left. Show all your work in the space provided.

6. A pure solid at a fixed temperature has a constant density. We know that

$$density = \frac{mass}{volume} \text{ or } D = \frac{m}{V}.$$

_____directly_____ **a.** Are mass and volume directly proportional or inversely proportional for a fixed density?

_____6.0cm³_____ **b.** If a solid has a density of 4.0 g/cm³, what volume provides 24 g of that solid?

7. An adding-machine tape has a width of 5.80 cm. A long strip of it is torn off and measured to be 5.6 m long.

_____560_____ **a.** Convert 5.6 m into centimeters.

_____3200cm³_____ **b.** What is the area of this rectangular strip of tape, in cm²?

MODERN CHEMISTRY

CHAPTER **2** REVIEW
Measurements and Calculations

MIXED REVIEW

SHORT ANSWER Answer the following questions in the space provided.

1. Match the description on the right to the appropriate quantity on the left.

_____ 2 m³   **(a)** mass of a small paper clip

_____ 0.5 g   **(b)** length of a small paper clip

_____ 0.5 kg   **(c)** length of a stretch limousine

_____ 600 cm²   **(d)** volume of a refrigerator compartment

_____ 20 mm   **(e)** surface area of the cover of this workbook

                              (f) mass of a jar of peanut butter

2. __b__ A measurement is said to have good precision if ____.

(a) it agrees closely with an accepted standard
(b) it agrees closely with other similar measurements
(c) it has a small number of significant figures
(d) it has a large number of significant figures

3. A certain sample with a mass of 4.00 g is found to have a volume of 7.0 mL. A student entered 4.00 ÷ 7.0 on a calculator. The calculator display shows the answer as 0.571429.

_____ *yes* _____ **a.** Is the setup for calculating density correct?

_____ *2* _____ **b.** How many significant figures should the answer contain?

4. It was shown in the text that in a value such as 4000 g, the precision of the number is uncertain. The zeros may or may not be significant.

_____ *1* _____ **a.** Suppose that the mass was determined to be 4000 g to the nearest gram. How many significant figures are present in this measurement?

_____ $4.00 \times 10^3 g$ _____ **b.** Suppose next you are told that the mass in fact lies somewhere between 3950 and 4050 g. Use scientific notation to report the average value, showing an appropriate number of significant figures.

5. If you divide a sample's mass by its density, what are the resulting units?

_____ *volume units for example* _____

6. Three students were asked to determine the volume of a liquid by a method of their choosing. Each did three trials. The table below shows the results. (The actual volume is 24.8 mL.)

	Trial 1 (mL)	Trial 2 (mL)	Trial 3 (mL)
Student A	24.2	24.6	24.0
Student B	24.2	24.3	24.3
Student C	24.7	24.8	24.7

_____C_____ **a.** Which students' measurements showed the greatest accuracy?

_____C, B_____ **b.** Which students' measurements showed the greatest precision?

PROBLEMS Write the answer on the line to the left. Show all your work in the space provided.

7. ___2.0×10^2 g___ A single atom of platinum has a mass of 3.25×10^{-22} g. What is the mass of 6.0×10^{23} platinum atoms?

8. A sample thought to be pure lead occupies a volume of 15.0 mL and has a mass of 160.0 g.

___10.7 g/mL___ **a.** Determine its density.

___NO___ **b.** Is the sample pure lead? (Refer to Table 2-4 on page 38 of the text.)

___5.7%___ **c.** Determine the percent error, based on the accepted density for lead.

Name ______________________ Date ______________ Class ______________

CHAPTER 4 REVIEW

Arrangement of Electrons in Atoms

SECTION 4-1

SHORT ANSWER **Answer the following questions in the space provided.**

1. How does the photoelectric effect support the particle theory of light?

2. What is the difference between the ground state and the excited state of electron positions?

3. How can an atom emit a photon?

4. How can the energy levels of electrons be determined by measuring the light emitted from an atom?

5. Why does electromagnetic radiation in the ultraviolet region represent a larger energy transition than does radiation in the infrared region?

Name _____ Date _____ Class _____

6. Which of the waves shown below has the higher frequency? Explain your answer.

_____

_____

_____

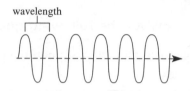

7. How many times were photons of radiation emitted from excited helium atoms to form the spectrum shown below? Explain your answer.

_____

_____

_____

_____

_____

Spectrum for helium

PROBLEMS Write the answer on the line to the left. Show all your work in the space provided.

8. _____ The wavelength of light is 310 nm. Calculate its frequency.

9. _____ What is the wavelength of electromagnetic radiation if its frequency is 3.2×10^{-2} Hz?

Name _____ Date _____ Class _____

SECTION 4-2

SHORT ANSWER Answer the following questions in the space provided.

1. __d__ How many quantum numbers are used to describe the energy state of an electron in an atom?

    (**a**) 1   (**c**) 3

    (**b**) 2   (**d**) 4

2. __a__ A spherical electron cloud surrounding an atomic nucleus would best represent _____ .

    (**a**) an *s* orbital   (**c**) a combination of two different *p* orbitals

    (**b**) a *p* orbital   (**d**) a combination of an *s* and a *p* orbital

3. __C__ An energy level of $n = 4$ can hold _____ electrons.

    (**a**) 32   (**c**) 8

    (**b**) 24   (**d**) 6

4. __C__ An energy level of $n = 2$ can hold _____ electrons.

    (**a**) 32   (**c**) 8

    (**b**) 24   (**d**) 6

5. __b__ An electron for which $n = 4$ has more _____ than an electron for which $n = 2$.

    (**a**) spin   (**c**) energy

    (**b**) stability   (**d**) wave nature

6. _____ According to Bohr, electrons cannot reside at _____ in the figure below.

    (**a**) point A   (**c**) point C

    (**b**) point B   (**d**) point D

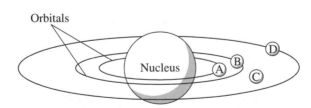

7. __C__ According to the quantum theory, point D in the above figure represents _____.

    (**a**) the fixed position of an electron

    (**b**) the farthest position from the nucleus that an electron can achieve

    (**c**) a position where an electron probably exists

    (**d**) a position where an electron cannot exist

SECTION 4-2 continued

8. How did de Broglie conclude that electrons have a wave nature?

_____

_____

_____

_____

_____

_____

_____

_____

_____

9. Identify each of the four quantum numbers and the properties to which they refer.

_____

_____

_____

_____

_____

_____

10. How did the Heisenberg uncertainty principle contribute to the idea that electrons occupy "clouds," or "orbitals"?

_____

_____

_____

11. Complete the following table.

Principal quantum number, n	Number of sublevels	Types of orbitals
1		
2		
3		
4		

Name ______________________ Date ______________ Class ______________

CHAPTER 4 REVIEW

Arrangement of Electrons in Atoms

SECTION 4-3

SHORT ANSWER **Answer the following questions in the space provided.**

1. Compare and contrast Hund's rule with the Pauli exclusion principle.

____

____

____

____

2. Explain the conditions under which the following orbital notation for helium is possible:

↑	↑
1*s*	2*s*

____

____

Write the electron configuration and orbital notation for each of the following atoms.

3. Phosphorus

4. Nitrogen

5. Potassium

SECTION 4-3 continued

6. Aluminum

7. Argon

8. Boron

9. Which guideline, Hund's rule or the Pauli exclusion principle, is violated in the following orbital diagrams?

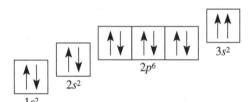

a. _____

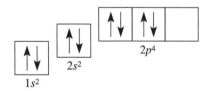

b. _____

MODERN CHEMISTRY

CHAPTER 4 REVIEW

Arrangement of Electrons in Atoms

MIXED REVIEW

SHORT ANSWER Answer the following questions in the space provided.

1. Under what conditions does matter create light?

_____

_____

_____

2. What do quantum numbers describe?

_____

_____

_____

3. What is the relationship between the principal quantum number and the electron configuration?

_____

_____

_____

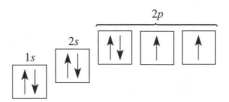

4. How does the figure above illustrate Hund's rule?

_____

_____

_____

5. How does the figure above illustrate the Pauli exclusion principle?

_____

_____

Name ______________________ Date ______________ Class ______________

MIXED REVIEW continued

6. Elements of the fourth and higher main-energy levels do not seem to follow the normal sequence for filling orbitals. Why is this so?

____

____

____

7. How do electrons create the colors in a line-emission spectrum?

____

____

____

8. Write the electron configuration of the following atoms:

a. carbon

____

b. potassium

____

c. gallium

____

d. copper

____

PROBLEM **Write the answer on the line to the left. Show all your work in the space provided.**

9. ______________ What would be the wavelength of light that has a frequency of 3×10^{-4} Hz in a vacuum?

| CHAPTER **5** REVIEW |

The Periodic Law

SECTION 5-1

SHORT ANSWER   Answer the following questions in the space provided.

1. _____ In the modern periodic table, elements are ordered _____.

    (a) according to decreasing atomic mass
    (b) according to Mendeleev's original design
    (c) according to increasing atomic number
    (d) based on when they were discovered

2. _____ Mendeleev noticed that properties of elements appeared at regular intervals when the elements were arranged in order of increasing _____.

    (a) density   **(c)** atomic number
    (b) reactivity   **(d)** atomic mass

3. _____ The modern periodic law states that _____.

    (a) no two electrons with the same spin can be found in the same place in an atom
    (b) the physical and chemical properties of the elements are functions of their atomic number
    (c) electrons exhibit properties of both particles and waves
    (d) the chemical properties of elements can be grouped according to periodicity, but physical properties cannot

4. _____ The discovery of the noble gases changed Mendeleev's periodic table by adding a new _____.

    (a) period   **(c)** group
    (b) series   **(d)** sublevel block

5. _____ The most distinctive property of the noble gases is that they are _____.

    (a) metallic   **(c)** metalloids
    (b) radioactive   **(d)** largely unreactive

6. _____ Lithium, the first element in Group 1, has an atomic number of 3. The second element in this group has an atomic number of _____.

    (a) 4   **(c)** 11
    (b) 10   **(d)** 18

7. An isotope of fluorine has a mass number of 19 and an atomic number of 9.

    _____ **a.** How many protons are in this atom?

    _____ **b.** How many neutrons are in this atom?

    _____ **c.** What is the symbol of this fluorine atom including its mass number and atomic number?

8. Samarium, Sm, is a member of the lanthanide series.

_____ **a.** Identify the element just below samarium in the periodic table.

_____ **b.** The atomic numbers of these two elements differ by how many units?

9. A certain isotope contains 53 protons, 78 neutrons, and 54 electrons.

_____ **a.** What is its atomic number?

_____ **b.** What is the mass of this atom in amus (to the nearest whole number)?

_____ **c.** Is this element Pt, Xe, I, or Bh?

_____ **d.** Identify two other elements that are in its group.

10. In a modern periodic table, every element is a member of both a horizontal row and a vertical column. Which one is the group, and which one is the period?

_____

_____

11. Explain the distinction between atomic mass and atomic number.

_____

_____

_____

_____

12. In the periodic table, the atomic masses of Te and I decrease rather than increase, while their atomic numbers increase. This phenomenon happens to other neighboring elements in the periodic table. Name two of these sets of elements.

_____

_____

CHAPTER **5** REVIEW
The Periodic Law

SECTION 5-2

SHORT ANSWER Answer the following questions in the space provided.

Use only this periodic table to answer the following questions.

E																	E

1 H

																	G 2 He

B 3 Li	C 4 Be												5 B	6 C	7 N	8 O	F 9 F	10 Ne
11 Na	12 Mg					D							13 Al	14 Si	15 P	16 S	17 Cl	18 Ar
19 K	20 Ca	21 Sc	22 Ti	23 V	24 Cr	25 Mn	26 Fe	27 Co	28 Ni	29 Cu	30 Zn	31 Ga	32 Ge	33 As	34 Se	35 Br	36 Kr
37 Rb	38 Sr	39 Y	40 Zr	41 Nb	42 Mo	43 Tc	44 Ru	45 Rh	46 Pd	47 Ag	48 Cd	49 In	50 Sn	51 Sb	52 Te	53 I	54 Xe
55 Cs	56 Ba	57 La	72 Hf	73 Ta	74 W	75 Re	76 Os	77 Ir	78 Pt	79 Au	80 Hg	81 Tl	82 Pb	83 Bi	84 Po	85 At	86 Rn
87 Fr	88 Ra	89 Ac	104 Rf	105 Db	106 Sg	107 Bh	108 Hs	109 Mt									

58 Ce	59 Pr	60 Nd	61 Pm	62 Sm	63 Eu	64 Gd	65 Tb	66 Dy	67 Ho	68 Er	69 Tm	70 Yb	71 Lu
90 Th	91 Pa	92 U	93 Np	94 Pu	95 Am	96 Cm	97 Bk	98 Cf	99 Es	100 Fm	101 Md	102 No	103 Lr

H

1. Identify the element, and write the noble-gas notation for each of the following:

   a. the Group 14 element in Period 4

   _____

   b. the only metal in Group 15

   _____

   c. the transition metal with the smallest atomic mass

   _____

   d. the alkaline earth metal with the largest atomic number

   _____

SECTION 5-2 continued

2. On the periodic table at the top of page 35 several areas are labeled A–H.

_____ **a.** Area A represents which block, *s, p, d,* or *f?*

   b. Identify the remaining labeled areas of the table, choosing from the following terms: main-group elements, transition elements, lanthanides, actinides, alkali metals, alkaline-earth metals, halogens, noble gases.

   _____ B

   _____ C

   _____ D

   _____ E

   _____ F

   _____ G

   _____ H

3. Give the symbol, period, group, and block for the following:

   a. sulfur

   _____

   _____

   b. nickel

   _____

   _____

   c. $[Kr]5s^1$

   _____

   _____

   d. $[Ar]3d^54s^1$

   _____

   _____

4. There are 18 columns in the periodic table; each has a group number. Give the group numbers that make up each of the following blocks:

   _____ **a.** *s* block

   _____ **b.** *p* block

   _____ **c.** *d* block

CHAPTER 5 REVIEW
The Periodic Law

SECTION 5-3

SHORT ANSWER   Answer the following questions in the space provided.

1. _____ When an electron is added to a neutral atom, energy is ____.

     (a) always absorbed   **(c)** either absorbed or released
     (b) always released   **(d)** burned away

2. _____ The energy required to remove an electron from an atom is the atom's ____.

     (a) electron affinity   **(c)** electronegativity
     (b) electron energy   **(d)** ionization energy

3. Moving from left to right across a period on the periodic table,

     _____ **a.** electron affinity values tend to become ____ (more negative or more positive).

     _____ **b.** ionization energy values tend to become ____ (larger or smaller).

     _____ **c.** atomic radii tend to become ____ (larger or smaller).

4. _____ **a.** Name the halogen with the least-negative electron affinity.

     _____ **b.** Name the alkali metal with the highest ionization energy.

     _____ **c.** Name the element in Period 3 with the smallest atomic radius.

     _____ **d.** Name the Group 14 element with the largest electronegativity.

5. Write the electron configuration of the following:

     a. Na

     _____

     b. Na^+

     _____

     c. O

     _____

     d. O^{2-}

     _____

     e. Co^{2+}

     _____

SECTION 5-3 continued

6. a. Compare the size of the radius of a positive ion to its neutral atom.

_____

_____

b. Compare the size of the radius of a negative ion to its neutral atom.

_____

_____

7. a. Give the approximate positions and blocks where metals and nonmetals reside in the periodic table.

_____

_____

_____

b. Of metals and nonmetals, which tend to form positive ions? Which tend to form negative ions?

_____

_____

8. Table 5-3 on page 145 of the text lists successive ionization energies for several elements.

_____ **a.** Identify the electron that is removed in the first ionization energy of Mg.

_____ **b.** Identify the electron that is removed in its second ionization energy.

_____ **c.** Identify the electron that is removed in its third ionization energy.

d. Explain why the second ionization energy is higher than the first, the third is higher than the second, and so on.

_____

_____

_____

_____

9. Explain the role of valence electrons in the formation of chemical compounds.

_____

_____

_____

_____

CHAPTER 6 REVIEW
Chemical Bonding

SECTION 6-1

SHORT ANSWER Answer the following questions in the space provided.

1. _____ A chemical bond between atoms results from the attraction between electrons and _____.

    (a) protons   **(c)** isotopes

    (b) neutrons   **(d)** Lewis structures

2. _____ A covalent bond consists of _____.

    (a) a shared electron   **(c)** two different ions

    (b) a shared electron pair   **(d)** an octet of electrons

3. _____ If two covalently bonded atoms are identical, the bond is identified as _____.

    (a) nonpolar covalent   **(c)** nonionic

    (b) polar covalent   **(d)** dipolar

4. _____ A covalent bond in which there is an unequal attraction for the shared electrons is _____.

    (a) nonpolar   **(c)** ionic

    (b) polar   **(d)** dipolar

5. _____ Atoms with a strong attraction for electrons they share with another atom exhibit _____.

    (a) zero electronegativity   **(c)** high electronegativity

    (b) low electronegativity   **(d)** Lewis electronegativity

6. _____ Bonds that possess between 5% and 50% ionic character are considered to be _____.

    (a) ionic   **(c)** polar covalent

    (b) pure covalent   **(d)** nonpolar covalent

7. _____ The greater the electronegativity difference between two atoms bonded together, the greater the bond's percentage of _____.

    (a) ionic character   **(c)** metallic character

    (b) covalent character   **(d)** electron sharing

8. The electrons involved in the formation of a chemical bond are called

_____.

9. A chemical bond that results from the electrostatic attraction between positive and

negative ions is called a(n) _____.

SECTION 6-1 continued

10. If electrons involved in bonding spend most of the time closer to one atom rather than

the other, the bond is _____.

11. If a bond's character is more than 50% ionic, then the bond is called

a(n) _____.

12. A bond's character is more than 50% ionic if the electronegativity difference between the two

atoms is greater than _____.

13. Write the formula for an example of each of the following compounds:

_____ **a.** nonpolar covalent compound

_____ **b.** polar covalent compound

_____ **c.** ionic compound

14. Describe how a covalent bond holds two atoms together.

_____

_____

_____

15. What property of the two atoms in a covalent bond determines whether or not the bond will be
polar?

_____

_____

16. How can electronegativity be used to distinguish between an ionic bond and a covalent bond?

_____

_____

_____

_____

17. What is unique about the bonding properties of carbon?

_____

_____

_____

CHAPTER 6 REVIEW
Chemical Bonding

SECTION 6-2

SHORT ANSWER Answer the following questions in the space provided.

1. Use the concept of potential energy to describe how a covalent bond forms between two atoms.

_____

_____

_____

_____

_____

_____

2. Name two elements that form compounds that are exceptions to the octet rule.

_____

_____

3. Explain why resonance structures illustrate the limitations of Lewis structures in correctly modeling covalent bonds.

_____

_____

_____

_____

4. Bond energy is related to bond length. Use the data in the tables below to arrange the bonds listed in order of increasing bond length, from shortest bond to longest.

a.

Bond	Bond energy (kJ/mol)
H—F	569
H—I	299
H—Cl	432
H—Br	366

_____

b.

Bond	Bond energy (kJ/mol)
C—C	346
C≡C	835
C=C	612

_____

_____

5. Draw Lewis structures to represent each of the following formulas:

a. NH_3

b. H_2O

c. CH_4

d. C_2H_2

e. CH_2O

CHAPTER 6 REVIEW
Chemical Bonding

SECTION 6-3

SHORT ANSWER   Answer the following questions in the space provided.

1. _____ The notation for sodium chloride, NaCl, stands for one _____.

    (a) formula unit   **(c)** crystal

    (b) molecule   **(d)** atom

2. _____ In a crystal of an ionic compound, each cation is surrounded by a number of _____.

    (a) molecules   **(c)** dipoles

    (b) positive ions   **(d)** negative ions

3. _____ Compared with the neutral atoms involved in the formation of an ionic compound, the crystal lattice that results is _____.

    (a) higher in potential energy   **(c)** equal in potential energy

    (b) lower in potential energy   **(d)** unstable

4. _____ The lattice energy of compound A is greater than that of compound B. What can be concluded from this fact?

    (a) Compound A is not an ionic compound.

    (b) It will be more difficult to break the bonds in compound A than in compound B.

    (c) Compound B is probably a gas.

    (d) Compound A has larger crystals than compound B.

5. _____ The forces of attraction between molecules in a molecular compound are _____.

    (a) stronger than the attractive forces in ionic bonding

    (b) weaker than the attractive forces in ionic bonding

    (c) approximately equal to the attractive forces in ionic bonding

    (d) equal to zero

6. Describe the force that holds two atoms together in an ionic bond.

_____

_____

_____

7. What type of energy best represents the strength of an ionic bond?

_____

_____

SECTION 6-3 continued

8. What type of bonding holds a polyatomic ion together?

____

____

____

____

9. Arrange the ionic bonds in the table below in order of increasing strength from weakest bond to strongest.

Ionic bond	Lattice energy (kJ/mol)
NaCl	−787
CaO	−3384
KCl	−715
MgO	−3760
LiCl	−861

____

____

10. Draw Lewis structures for the following polyatomic ions:

a. NH_4^+

b. SO_4^{2-}

11. Draw the two resonance structures for the nitrite anion, NO_2^-.

CHAPTER 6 REVIEW
Chemical Bonding

SECTION 6-4

SHORT ANSWER Answer the following questions in the space provided.

1. _____ In metals, the valence electrons are considered to be _____.

    (a) attached to particular positive ions   **(c)** immobile

    (b) shared by all surrounding atoms   **(d)** involved in covalent bonds

2. _____ The fact that metals are malleable and ionic crystals are brittle is best explained in terms of their _____.

    (a) chemical bonds   **(c)** heats of vaporization

    (b) London forces   **(d)** polarity

3. _____ As light strikes the surface of a metal, the electrons in the electron sea _____.

    (a) allow the light to pass through

    (b) become attached to particular positive ions

    (c) fall to lower energy levels

    (d) absorb and re-emit the light

4. _____ Mobile electrons in the metallic bond are responsible for _____.

    (a) luster   **(c)** electrical conductivity

    (b) thermal conductivity   **(d)** all of the above

5. _____ In general, the strength of the metallic bond _____ moving from left to right on any row of the periodic table.

    (a) increases

    (b) decreases

    (c) remains the same

6. _____ When a metal is drawn into a wire, the metallic bonds _____.

    (a) break easily   **(c)** do not break

    (b) break with difficulty   **(d)** become ionic bonds

7. Use the concept of electron configurations to explain why the number of valence electrons in metals tends to be less than the number in most nonmetals.

_____

_____

_____

_____

8. How does the behavior of electrons in metals contribute to the metal's ability to conduct electricity and heat?

_____

_____

_____

9. What is the relationship between the heat of vaporization of a metal and the strength of the bonds that hold the metal together?

_____

_____

_____

10. Draw two diagrams of a metallic bond. In the first diagram, draw a weak metallic bond; in the second, show a metallic bond that would be stronger. Be sure to include nuclear charge and number of electrons in your illustrations.

a.   b.

weak bond   strong bond

11. Complete the following table:

	Metallic	Ionic
Components		
Overall charge		
Conductivity		
Melting point		
Hardness		
Malleable		
Ductile		

MODERN CHEMISTRY

CHAPTER 6 REVIEW
Chemical Bonding

SECTION 6-5

SHORT ANSWER   Answer the following questions in the space provided.

1. Identify the major assumption of the VSEPR theory that is used to predict the shape of atoms.

_____

_____

2. In water, two hydrogen atoms are bonded to one oxygen atom. Why isn't water a linear molecule?

_____

_____

3. What orbitals combine together to form sp^3 hybrid orbitals around a carbon atom?

_____

_____

4. What two factors determine whether or not a molecule is polar?

_____

_____

5. Arrange the following types of attractions in order of increasing strength, with 1 being the weakest and 4 the strongest.

_____ covalent

_____ ionic

_____ dipole-dipole

_____ London dispersion

6. How are dipole-dipole attractions, London dispersion forces, and hydrogen bonding similar?

_____

_____

_____

_____

SECTION 6-5 continued

7. Complete the following table:

Formula	Lewis structure	Geometry	Polar
H_2S			
CCl_4			
BF_3			
H_2O			
PCl_5			
BeF_2			
SF_6			

MODERN CHEMISTRY

CHAPTER 6 REVIEW
Chemical Bonding

MIXED REVIEW

SHORT ANSWER   Answer the following questions in the space provided.

1. Name the type of energy that is a measure of strength for each of the following types of bonds:

   _____ **a.** ionic bond

   _____ **b.** covalent bond

   _____ **c.** metallic bond

2. Use the electronegativity values shown in Figure 5-20, on page 151 of the text, to determine whether each of the following bonds is nonpolar covalent, polar covalent, or ionic.

   _____ **a.** H—F   _____ **d.** H—H

   _____ **b.** Na—Cl   _____ **e.** H—C

   _____ **c.** H—O   _____ **f.** H—N

3. How is a hydrogen bond different from an ionic or covalent bond?

   _____

   _____

   _____

   _____

   _____

   _____

4. H_2S and H_2O have similar structures and their central atoms belong to the same group. Yet H_2S is a gas at room temperature and H_2O is a liquid. Use bonding principles to explain why this is true.

   _____

   _____

   _____

   _____

   _____

MIXED REVIEW continued

5. Why is a polar-covalent bond similar to an ionic bond?

____

____

____

____

6. Draw a Lewis structure for each of the following formulas. Determine whether the molecule is polar or nonpolar.

________________ **a.** H_2S

________________ **b.** $COCl_2$

________________ **c.** PCl_3

________________ **d.** CH_2O

CHAPTER 7 REVIEW

Chemical Formulas and Chemical Compounds

SECTION 7-1

SHORT ANSWER Answer the following questions in the space provided.

1. _____ In a Stock name such as iron(III) sulfate, the roman numeral tells us _____.

    (a) how many atoms of Fe are in one formula unit
    (b) how many sulfate ions can be attached to the iron atom
    (c) the charge on each Fe ion
    (d) the total positive charge of the formula unit

2. _____ The result of changing a subscript in a correctly written chemical formula is to _____.

    (a) change the number of moles represented by the formula
    (b) change the charges on the other ions in the compound
    (c) change the formula so that it no longer represents the compound it previously
    represented
    (d) have no effect on the formula

3. The explosive TNT has the molecular formula $C_7H_5(NO_2)_3$.

_____ **a.** How many elements make up this compound?

_____ **b.** How many oxygen atoms are present in one molecule of $C_7H_5(NO_2)_3$?

_____ **c.** How many atoms in total are present in one molecule of $C_7H_5(NO_2)_3$?

_____ **d.** How many atoms are present in a sample of 2×10^{23} molecules of $C_7H_5(NO_2)_3$?

4. How many atoms are present in each of these formula units?

_____ **a.** $Ca(HCO_3)_2$

_____ **b.** $C_{12}H_{22}O_{11}$

_____ **c.** $Fe(ClO_2)_3$

_____ **d.** $Fe(ClO_3)_2$

5. _____ **a.** What is the formula for the compound dinitrogen pentoxide?

_____ **b.** What is the Stock name for the covalent compound CS_2?

_____ **c.** What is the formula for sulfurous acid?

_____ **d.** What is the name for the acid H_3PO_4?

6. Some binary compounds are ionic, others are covalent. The types of bonding partially depend on the position of the elements in the periodic table. Label each of these claims as True or False; if False, specify the nature of the error.

    a. Covalently bonded binary molecular compounds typically form from nonmetals.

    _____

    _____

    b. Binary ionic compounds form from metals and nonmetals, typically from opposite sides of the periodic table.

    _____

    _____

    c. Binary compounds involving metalloids are always ionic.

    _____

    _____

7. Refer to Table 7-2 on page 210 of the text and Table 7-5 on page 214 of the text for examples of names and formulas for polyatomic ions and acids.

    a. Derive a generalization for when an acid name will end in the suffix *-ic* or *-ous*.

    _____

    _____

    _____

    b. Derive a generalization for when an acid name will begin with the prefix *hydro-* and when it will not.

    _____

    _____

    _____

8. Fill in the blanks in the table below.

Compound name	Formula
Aluminum sulfide	
Aluminum sulfite	
	$PbCl_2$
	$(NH_4)_3PO_4$
Hydroiodic acid	

Chemical Formulas and Chemical Compounds

SECTION 7-2

SHORT ANSWER   Answer the following questions in the space provided.

1. Assign the oxidation number to the specified element in each of the following examples:

_____   **a.** S in H_2SO_3

_____   **b.** S in $MgSO_4$

_____   **c.** S in K_2S

_____   **d.** Cu in Cu_2S

_____   **e.** Cr in Na_2CrO_4

_____   **f.** H in $(HCO_3)^-$

_____   **g.** C in $(HCO_3)^-$

_____   **h.** N in $(NH_4)^+$

2. _____ **a.** What is the formula for the compound sulfur(II) chloride?

_____ **b.** What is the Stock name for NO_2?

3. _____ **a.** Use electronegativity values to determine the one element that always has a negative oxidation number when it appears in any binary compound.

_____ **b.** What is the oxidation number and formula for the element described in part a when it exists as an uncombined element?

4. Tin has possible oxidation numbers of $+2$ and $+4$ and forms two known oxides. One of them has the formula SnO_2.

_____ **a.** Give the Stock name for SnO_2.

_____ **b.** Give the empirical formula for the other oxide of tin.

5. Scientists believe that two separate reactions contribute to the depletion of the ozone layer, O_3. The first reaction involves oxides of nitrogen. The second involves free chlorine atoms. Both reactions follow. When a compound is not stated as a formula, write the correct formula in the blank beside its name.

Oxides of nitrogen:

a. _____ (nitrogen monoxide) $+ O_3 \rightarrow$ _____ (nitrogen dioxide) $+ O_2$

Free chlorine:

b. $Cl + O_3 \rightarrow$ _____ (chlorine monoxide) $+ O_2$

6. Consider the covalent compound dinitrogen trioxide when answering the following:

_____ **a.** What is the formula for dinitrogen trioxide?

_____ **b.** What is the oxidation number assigned to each N atom in this compound? Explain your answer.

_____

_____

_____

_____ **c.** Give the Stock name for dinitrogen trioxide.

7. The oxidation numbers assigned to the atoms in some organic compounds sometimes give unexpected results. Assign oxidation numbers to each atom in the following compounds:

a. CO_2

_____

b. CH_4 (methane)

_____

c. $C_6H_{12}O_6$ (glucose)

_____

d. C_3H_8 (propane gas)

_____

8. Assign oxidation numbers to each element in the compounds found in the following situations:

a. Rust, Fe_2O_3, forms on an old nail.

_____

b. Nitrogen dioxide, NO_2, pollutes the air as smog.

_____

c. Chromium dioxide, CrO_2, is used to make recording tapes.

_____

CHAPTER 7 REVIEW

Chemical Formulas and Chemical Compounds

SECTION 7-3

SHORT ANSWER   Answer the following questions in the space provided.

1. Label each of the following statements as True or False:

_____ **a.** If the formula mass of one molecule is x amu, the molar mass is x g/mol.

_____ **b.** Samples of two different chemicals with equal numbers of moles must have equal masses as well.

_____ **c.** Samples of two different chemicals with equal numbers of moles must have equal numbers of molecules as well.

2. How many moles of each element are present in a 10.0 mol sample of $Ca(NO_3)_2$?

_____

PROBLEMS   Write the answer on the line to the left. Show all your work in the space provided.

3. Consider a sample of 10.0 g of the gaseous hydrocarbon C_3H_4 to answer the following questions.

_____ **a.** How many moles are present in this sample?

_____ **b.** How many molecules are present in the C_3H_4 sample?

_____ **c.** How many carbon atoms are present in this sample?

_____ **d.** What is the percentage composition of hydrogen in the sample?

4. The chief source of aluminum metal is the ore alumina, Al_2O_3.

_____ **a.** Determine the percentage composition of Al in this ore.

_____ **b.** How many pounds of aluminum can be extracted from 2.0 tons of alumina?

5. Compound A has a molar mass of 20 g/mol, and compound B has a molar mass of 30 g/mol.

_____ **a.** What is the mass of 1.0 mol of compound A, in grams?

_____ **b.** How many moles are present in 5.0 g of compound B?

_____ **c.** How many moles of compound B are needed to have the same mass as 6.0 mol of compound A?

CHAPTER 7 REVIEW

Chemical Formulas and Chemical Compounds

SECTION 7-4

SHORT ANSWER Answer the following questions in the space provided.

1. Write empirical formulas to match the following molecular formulas:

_____ **a.** $C_2H_6O_4$

_____ **b.** N_2O_5

_____ **c.** Hg_2Cl_2

_____ **d.** C_6H_{12}

2. _____ A certain hydrocarbon has an empirical formula of CH_2 and a molar mass of 56.12 g/mol. What is its molecular formula?

3. A certain ionic compound is found to contain 0.012 mol of sodium, 0.012 mol of sulfur, and 0.018 mol of oxygen.

_____ **a.** What is its empirical formula?

_____ **b.** Is this compound a sulfate, sulfite, or neither?

PROBLEMS Write the answer on the line to the left. Show all your work in the space provided.

4. Water of hydration was discussed in Sample Problem 7-11 on pages 227–228 of the text. Strong heating will drive off the water as a vapor in hydrated copper(II) sulfate. Use the data table below to answer the following:

Mass of the empty crucible	4.00 g
Mass of the crucible plus hydrate sample	4.50 g
Mass of the system after heating	4.32 g
Mass of the system after a second heating	4.32 g

_____ **a.** Determine the percent water of hydration in the original sample.

_____ **b.** The compound has the formula $CuSO_4 \cdot xH_2O$. Determine value of x.

c. What might be the purpose of the second heating?

_____

_____

_____

_____

5. Gas X is found to be 24.0% carbon and 76.0% fluorine by mass.

_____ **a.** Determine the empirical formula of gas X.

_____ **b.** Given that the molar mass of gas X is 200.04 g/mol, determine its molecular formula.

6. A compound is found to contain 43.2% copper, 24.1% chlorine, and 32.7% oxygen by mass.

_____ **a.** Determine its empirical formula.

b. What is the correct Stock name of the compound in part a?

_____

Name _____ Date _____ Class _____

Chemical Formulas and Chemical Compounds

MIXED REVIEW

SHORT ANSWER Answer the following questions in the space provided.

1. Write formulas for the following compounds:

_____ **a.** copper(II) carbonate

_____ **b.** sodium sulfite

_____ **c.** ammonium phosphate

_____ **d.** tin(IV) sulfide

_____ **e.** nitrous acid

2. Write the Stock names for the following compounds:

_____ **a.** $Mg(ClO_4)_2$

_____ **b.** $Fe(NO_3)_2$

_____ **c.** $Fe(NO_2)_3$

_____ **d.** CoO

_____ **e.** dinitrogen pentoxide

3. _____ **a.** How many atoms are represented by the formula $Ca(HSO_4)_2$?

_____ **b.** How many moles of oxygen atoms are in a 0.50 mol sample of this compound?

_____ **c.** Assign the oxidation number to sulfur in the HSO_4^- anion.

4. Assign the oxidation number to the element specified in each of the following:

_____ **a.** hydrogen in H_2O_2

_____ **b.** hydrogen in MgH_2

_____ **c.** sulfur in S_8

_____ **d.** carbon in $(CO_3)^{2-}$

_____ **e.** chromium in $Na_2Cr_2O_7$

_____ **f.** nitrogen in NO_2

HRW material copyrighted under notice appearing earlier in this work.

MIXED REVIEW continued

PROBLEMS   Write the answer on the line to the left. Show all your work in the space provided.

5. _____ Following are samples of four different compounds. Arrange them in order of increasing mass, from smallest to largest.

   a. 25 g of oxygen gas   **c.** 3×10^{23} molecules of C_2H_6
   b. 1.00 mol of H_2O **d.** 2×10^{23} molecules of $C_2H_6O_2$

6. _____   **a.** What is the formula for sodium hydroxide?

_____   **b.** What is the formula mass of sodium hydroxide?

_____   **c.** What is the mass in grams of 0.25 mol of sodium hydroxide?

7. _____ What is the percentage composition of ethane gas, C_2H_6, to the nearest whole number?

8. _____ Ribose is an important sugar (part of RNA), with a molar mass of 150.15 g/mol. If its empirical formula is CH_2O, what is its molecular formula?

MIXED REVIEW continued

9. Butane gas, C_4H_{10}, is often used as a fuel.

_____ **a.** What is the mass in grams of 3.00 mol of butane?

_____ **b.** How many molecules are present in that 3.00 mol sample?

_____ **c.** What is the empirical formula of the gas?

10. _____ Naphthalene is a soft covalent solid that is often used in mothballs. Its molar mass is 128.18 g/mol and it contains 93.75% carbon and 6.25% hydrogen. Determine the molecular formula of napthalene from this data.

11. Nicotine has the formula $C_xH_yN_z$. To determine its composition, a sample is burned in excess oxygen, producing the following results:

1.0 mol of CO_2
0.70 mol of H_2O
0.20 mol of NO_2

Assume that all the atoms in nicotine are present as products.

_____ **a.** Determine the number of moles of carbon present in the products of this combustion.

_____ **b.** Determine the number of moles of hydrogen present in the combustion products.

_____ **c.** Determine the number of moles of nitrogen present in the combustion products.

_____ **d.** Determine the empirical formula of nicotine based on your calculations.

_____ **e.** In a separate experiment, the molar mass of nicotine is found to be somewhere between 150 and 180 g/mol. Calculate the molar mass of nicotine to the nearest gram.

12. When $MgCO_3(s)$ is strongly heated, it produces solid MgO as gaseous CO_2 is driven off.

_____ **a.** What is the percentage loss in mass as this reaction occurs?

_____ **b.** Assign the oxidation number to each atom in $MgCO_3$?

_____ **c.** Does the oxidation number of carbon change upon forming CO_2?

CHAPTER 8 REVIEW
Chemical Equations and Reactions

SECTION 8-1

SHORT ANSWER Answer the following questions in the space provided.

1. Match the symbol on the left with its appropriate description on the right.

_____ Δ   (a) A precipitate forms.

_____ ↓   (b) A gas forms.

_____ ↑   (c) A reversible reaction occurs.

_____ (*l*)   (d) Heat is applied to the reactants.

_____ (*aq*)   (e) A chemical is dissolved in water.

_____ ⇌   (f) A chemical is in the liquid state.

2. Finish balancing the following equation:

$3Fe_3O_4 +$ _____ $Al \rightarrow$ _____ $Al_2O_3 +$ _____ Fe

3. In each of the following formulas with coefficients, write the total number of atoms present.

_____ **a.** $4SO_2$

_____ **b.** $8O_2$

_____ **c.** $3Al_2(SO_4)_3$

_____ **d.** $6 \times 10^{23}\ HNO_3$

4. Convert the following word equation into a balanced chemical equation:
aluminum metal + copper(II) fluoride → aluminum fluoride + copper metal

_____

5. One way to test the salinity of a water supply is to add a few drops of silver nitrate of known concentration to the water. As the solutions of sodium chloride and silver nitrate mix, a precipitate of silver chloride forms, leaving sodium nitrate in solution. Translate these sentences into a balanced chemical equation.

_____

6. a. Balance the following equation: $NaHCO_3(s) \xrightarrow{\Delta} Na_2CO_3(s) + H_2O(g) + CO_2(g)$

_____

SECTION 8-1 continued

    b. Translate the chemical equation in part a into a sentence.

_____

_____

_____

_____

7. The poisonous gas hydrogen sulfide can be neutralized with a base such as NaOH. The unbalanced equation for this reaction follows:

$$NaOH(aq) + H_2S(g) \rightarrow Na_2S(aq) + H_2O(l)$$

A student who was asked to balance this equation wrote the following:

$$Na_2OH(aq) + H_2S(g) \rightarrow Na_2S(aq) + H_3O(l)$$

Is this equation balanced? Is it correct? Explain why or why not, and supply the correct balanced equation if necessary.

_____

_____

_____

_____

PROBLEM Write the answer on the line to the left. Show all your work in the space provided.

8. Recall that coefficients in a balanced equation give relative amounts of moles as well as numbers of molecules.

_____ **a.** Calculate the amount of CO_2 in moles that forms if 10 mol of C_3H_4 react according to the following balanced equation:

$$C_3H_4 + 4O_2 \rightarrow 3CO_2 + 2H_2O$$

_____ **b.** Calculate the amount of O_2 in moles that is consumed.

MODERN CHEMISTRY

CHAPTER 8 REVIEW
Chemical Equations and Reactions

SECTION 8-2

SHORT ANSWER   Answer the following questions in the space provided.

1. Match the equation type on the left to its representation on the right.

_____ synthesis   **(a)** $AX + BY \rightarrow AY + BX$

_____ decomposition   **(b)** $A + BX \rightarrow AX + B$

_____ single replacement   **(c)** $A + B \rightarrow AX$

_____ double replacement   **(d)** $AX \rightarrow A + X$

2. _____ In the equation $2Al(s) + 3Fe(NO_3)_2(aq) \rightarrow 3Fe(s) + 2Al(NO_3)_3(aq)$, iron has been replaced by _____.

(a) nitrate   **(c)** aluminum
(b) water   **(d)** nitrogen

3. _____ Of the following chemical equations, the only reaction that is both synthesis and combustion is _____.

(a) $C(s) + O_2(g) \rightarrow CO_2(g)$
(b) $2C_4H_{10}(l) + 13O_2(g) \rightarrow 8CO_2(g) + 10H_2O(l)$
(c) $6CO_2(g) + 6H_2O(g) \rightarrow C_6H_{12}O_6(aq) + 6O_2(g)$
(d) $C_6H_{12}O_6(aq) + 6O_2(g) \rightarrow 6CO_2(aq) + 6H_2O(l)$

4. _____ Of the following chemical equations, the only reaction that is both decomposition and combustion is _____.

(a) $C(s) + O_2(g) \rightarrow CO_2(g)$
(b) $2C_4H_{10}(l) + 13O_2(g) \rightarrow 8CO_2(g) + 10H_2O(l)$
(c) $2H_2O_2(l) \rightarrow 2H_2O(l) + O_2(g)$
(d) $2HgO(s) \xrightarrow{\Delta} 2Hg(l) + O_2(g)$

5. Identify the products when the following substances decompose:

_____ **a.** a binary compound

_____ **b.** a metallic hydroxide

_____ **c.** a metallic carbonate

_____ **d.** the acid H_2SO_3

6. The complete combustion of a hydrocarbon in excess oxygen yields the products _____ and _____.

SECTION 8-2 continued

7. For the following four reactions, label the type, predict the products (make sure formulas are correct), and balance the equations:

   a. $Cl_2(aq) + NaI(aq) \rightarrow$

   _____

   _____

   b. $Mg(s) + N_2(g) \rightarrow$

   _____

   _____

   c. $Co(NO_3)_2(aq) + H_2S(aq) \rightarrow$

   _____

   _____

   d. $C_2H_5OH(aq) + O_2(g) \rightarrow$

   _____

   _____

8. Acetylene gas, C_2H_2, is burned to provide the high temperature needed in welding.

   a. Write the balanced chemical equation for the combustion of C_2H_2 in oxygen.

   _____

   _____

   _____ **b.** If 1.0 mol of C_2H_2 is burned, how many moles of CO_2 are formed?

   _____ **c.** How many moles of oxygen gas are consumed?

9. Write the balanced chemical equation for the reaction that occurs when solutions of barium chloride and sodium carbonate are mixed.

   _____

   _____

10. For the commercial preparation of aluminum metal, the metal is extracted from its ore, alumina, Al_2O_3 by electrolysis. Write the balanced chemical equation for the electrolysis of molten Al_2O_3.

   _____

   _____

CHAPTER 8 REVIEW
Chemical Equations and Reactions

SECTION 8-3

SHORT ANSWER Answer the following questions in the space provided.

1. List four metals that will *not* replace hydrogen in an acid.

_____

2. Consider the metals iron and silver, both listed in Table 8-3 on page 266 of the text. Which one readily forms oxides in nature, and which one does not?

_____

3. In each of the following pairs, identify the more-active element:

_____ **a.** F_2 and I_2

_____ **b.** Mn and K

_____ **c.** Cu and H

4. Use the information in Table 8-3 on page 266 of the text to predict whether each of the following reactions will occur. For those reactions that will occur, complete the chemical equation by writing in the products formed and balancing the final equation.

a. $Al(s) + CH_3COOH(aq) \xrightarrow{50°C}$

_____

_____

b. $Al(s) + H_2O(l) \xrightarrow{50°C}$

_____

_____

c. $Cr(s) + CdCl_2(aq) \rightarrow$

_____

_____

d. $Br_2(l) + KCl(aq) \rightarrow$

_____

_____

5. Very active metals will react with water to release hydrogen gas.

    a. Complete and then balance the equation for the reaction of Ca(*s*) with water.

_____

    b. The reaction of rubidium, Rb, with water is faster and more violent than the reaction of Na with water. Account for this difference in terms of the two metals' atomic structure and radius.

_____

_____

_____

_____

6. Gold is often used in jewelry. How does the relative activity of Au relate to its use in jewelry?

_____

_____

_____

7. Explain how to use an activity series to predict certain types of chemical behavior.

_____

_____

_____

_____

8. Aluminum is above copper in the activity series. Will aluminum metal react with copper(II) nitrate, $Cu(NO_3)_2$, to form aluminum nitrate, $Al(NO_3)_3$? If so, write the balanced equation for the reaction.

_____

_____

_____

_____

CHAPTER 8 REVIEW
Chemical Equations and Reactions

MIXED REVIEW

SHORT ANSWER Answer the following questions in the space provided.

1. _____ A balanced chemical equation represents all the following *except* _____.

   (a) experimentally established facts
   (b) the mechanism by which reactants recombine to form products
   (c) identities of reactants and products in a chemical reaction
   (d) relative quantities of reactants and products in a chemical reaction

2. _____ According to the law of conservation of mass, the total mass of the reacting substance is _____.

   (a) always more than the total mass of the products
   (b) always less than the total mass of the products
   (c) sometimes more and sometimes less than the total mass of the products
   (d) always equal to the total mass of the products

3. Predict whether each of the following chemical reactions will occur. For those reactions that will occur, label the reaction type and complete the chemical equation by writing in the products formed and balancing the final equation.

   a. $Ba(NO_3)_2(aq) + Na_3PO_4(aq) \rightarrow$

   _____

   _____

   b. $Al(s) + O_2(g) \rightarrow$

   _____

   _____

   c. $I_2(s) + NaBr(aq) \rightarrow$

   _____

   _____

   d. $C_3H_4(g) + O_2(g) \rightarrow$

   _____

   _____

   e. electrolysis of molten potassium chloride

_____

_____

4. Some small rockets are powered by the reaction shown in the following unbalanced equation:

$$(CH_3)_2N_2H_2(l) + N_2O_4(g) \rightarrow N_2(g) + H_2O(g) + CO_2(g) + heat$$

   a. Translate this chemical equation into a sentence. (Hint: The name for $(CH_3)_2N_2H_2$ is dimethylhydrazine.)

_____

_____

_____

   b. Balance the formula equation.

_____

_____

5. In the laboratory, you are given two small chips each of the unknown metals X, Y, and Z, along with dropper bottles containing solutions of $XCl_2(aq)$ and $ZCl_2(aq)$. Describe an experimental strategy you could use to determine the relative activities of X, Y, and Z.

_____

_____

_____

_____

_____

_____

_____

_____

_____

_____

6. List the observations that would indicate that a reaction has occurred.

_____

_____

CHAPTER 9 REVIEW
Stoichiometry

SECTION 9-1

SHORT ANSWER   Answer the following questions in the space provided.

1. _____ The coefficients in a chemical equation represent the _____.

   (a) masses in grams of all reactants and products
   (b) relative number of moles of reactants and products
   (c) number of atoms of each element in each compound in a reaction
   (d) number of valence electrons involved in a reaction

2. _____ Which of the following would not be studied within the topic of stoichiometry?

   (a) the mole ratio of Al to Cl in the compound aluminum chloride
   (b) the mass of carbon produced when a known mass of sucrose decomposes
   (c) the number of moles of hydrogen that will react with a known quantity of oxygen
   (d) the amount of energy required to break the ionic bonds in CaF_2

3. _____ A balanced chemical equation allows you to determine the _____.

   (a) mole ratio of any two substances in the reaction
   (b) energy released in the reaction
   (c) electron configuration of all elements in the reaction
   (d) reaction mechanism involved in the reaction

4. _____ The relative number of moles of hydrogen and oxygen that react to form water represents a(n) _____.

   (a) reaction sequence
   (b) bond energy
   (c) mole ratio
   (d) element proportion

5. Given the following unbalanced equation: $N_2O(g) + O_2(g) \rightarrow NO_2(g)$

   a. Balance the equation.

   _____

   _____ **b.** What is the mole ratio of NO_2 to O_2?

   _____ **c.** If 20.0 mol of NO_2 form, how many moles of O_2 must have been consumed?

   _____ **d.** Twice as many moles of NO_2 form as moles of N_2O are consumed. True or False?

   _____ **e.** Twice as many grams of NO_2 form as grams of N_2O are consumed. True or False?

PROBLEMS   Write the answer on the line to the left. Show all your work in the space provided.

6. Given the following equation: $N_2(g) + 3H_2(g) \rightarrow 2NH_3(g)$

_____ **a.** Determine to one decimal place the molar mass of each term, and write each one as a conversion factor from moles to grams.

_____

_____

b. There are three different mole ratios in this system. Write out each one.

_____

_____

7. Given the following equation: $4NH_3(g) + 6NO(g) \rightarrow 5N_2(g) + 6H_2O(g)$

_____ **a.** What is the mole ratio of NO to H_2O?

_____ **b.** What is the mole ratio of NO to NH_3?

_____ **c.** If 0.240 mol of NH_3 react according to the above equation, how many moles of NO will be consumed?

8. Propyne gas can be used as a fuel. The combustion reaction of propane can be represented by the following equation:

$$C_3H_4(g) + 4O_2(g) \rightarrow 3CO_2(g) + 2H_2O(g) + \text{heat energy}$$

a. Write all the possible mole ratios in this system.

_____

_____

_____

_____

b. Suppose that x moles of water form in the above reaction. The other three mole quantities (*not* in order) are $2x$, $1.5x$, and $0.5x$. Match these quantities to their respective formulas in the equation above.

_____

_____

CHAPTER 9 REVIEW
Stoichiometry

SECTION 9-2

PROBLEMS   Write the answer on the line to the left. Show all your work in the space provided.

1. _____ The following equation represents a laboratory preparation for oxygen gas:

$2KClO_3(s) + \text{heat} \rightarrow 2KCl(s) + 3O_2(g)$

How many moles of O_2 form as 3.0 mol of $KClO_3$ are totally consumed?

2. _____ Given the following equation: $H_2(g) + F_2(g) \rightarrow 2HF(g)$
How many grams of HF gas are produced as 5 mol of fluorine react?

3. _____ Water can be made to decompose into its elements by using electricity according to the following equation:

$2H_2O(l) + \text{electrical energy} \rightarrow 2H_2(g) + O_2(g)$

How many grams of O_2 are produced as 0.033 mol of water decompose?

4. _____ Sodium metal reacts with water to produce NaOH according to the following equation:

$2Na(s) + 2H_2O(l) \rightarrow 2NaOH(aq) + H_2(g)$

How many grams of NaOH are produced if 20.0 g of sodium metal react with excess oxygen?

---

5. _____ **a.** What mass of oxygen gas is produced if 100. g of lithium perchlorate are heated and allowed to decompose according to the following equation?

$$LiClO_4(s) \xrightarrow{\Delta} LiCl(s) + 2O_2(g)$$

_____ **b.** The oxygen gas produced in part a has a density of 1.43 g/L. Calculate the volume of the O_2 gas produced.

6. A car air bag requires 70. L of nitrogen gas to inflate properly. The following equation represents the production of nitrogen gas:

$$2NaN_3(s) \rightarrow 2Na(s) + 3N_2(g)$$

_____ **a.** The density of nitrogen gas is typically 1.16 g/L at room temperature. Calculate the number of grams of N_2 that are needed to inflate the air bag.

_____ **b.** Calculate the amount of N_2 in moles that are needed.

_____ **c.** Calculate the number of grams of NaN_3 that must be used to generate the amount of nitrogen gas necessary to properly inflate the air bag.

CHAPTER 9 REVIEW
Stoichiometry

SECTION 9-3

PROBLEMS   Write the answer on the line to the left. Show all your work in the space provided.

1. _____ If the actual yield of a reaction is 22 g and the theoretical yield is 25 g, calculate the percent yield.

2. 6.0 mol of N_2 are mixed with 12.0 mol of H_2 according to the following equation:

$$N_2(g) + 3H_2(g) \rightarrow 2NH_3(g)$$

_____ **a.** Which chemical is in excess? What is the excess amount in moles?

_____ **b.** Theoretically, how many moles of NH_3 will be produced?

_____ **c.** If the percent yield of NH_3 is 80%, how many moles of NH_3 are actually produced?

3. 0.050 mol of $Ca(OH)_2$ are combined with 0.080 mol of HCl according to the following equation:

$$Ca(OH)_2(aq) + 2HCl(aq) \rightarrow CaCl_2(aq) + 2H_2O(l)$$

_____ **a.** How many moles of HCl are required to neutralize all 0.050 mol of $Ca(OH)_2$?

_____ **b.** Which is the limiting reactant in this neutralization reaction?

_____ **c.** How many grams of water will form in this reaction?

4. Acid rain can form from the combustion of nitrogen gas producing $HNO_3(aq)$ in a two-step process.

$$N_2(g) + 2O_2(g) \rightarrow 2NO_2(g)$$

$$3NO_2(g) + H_2O(g) \rightarrow 2HNO_3(aq) + NO(g)$$

_____ **a.** A car burns 420. g of N_2 according to the above equations. How many grams of HNO_3 will be produced?

_____ **b.** For the above reactions to occur, O_2 must be in excess in the first step. What is the minimum amount of O_2 needed in grams?

_____ **c.** What volume does the amount of O_2 in part b occupy if its density is 1.4 g/L?

CHAPTER 9 REVIEW
Stoichiometry

MIXED REVIEW

SHORT ANSWER Answer the following questions in the space provided.

1. Given the following equation: $C_3H_4(g) + xO_2(g) \rightarrow 3CO_2(g) + 2H_2O(g)$

_____ **a.** What is the value of the coefficient x in this equation?

_____ **b.** What is the molar mass of C_3H_4?

_____ **c.** What is the mole ratio of O_2 to H_2O in the above equation?

_____ **d.** How many moles are in an 8.0 g sample of C_3H_4?

_____ **e.** If z mol of C_3H_4 react, how many moles of CO_2 are produced, in terms of z?

2. a. What is meant by "ideal conditions" relative to stoichiometric calculations?

_____

_____

b. What function do ideal stoichiometric calculations serve?

_____

_____

c. Are amounts actually produced typically larger or smaller than theoretical yields?

_____

PROBLEMS Write the answer on the line to the left. Show all your work in the space provided.

3. Assume the reaction represented by the following equation goes all the way to completion:

$$N_2 + 3H_2 \rightarrow 2NH_3$$

_____ **a.** If 6 mol of H_2 are consumed, how many moles of NH_3 are produced?

_____ **b.** How many grams are in a sample of NH_3 that contains 3.0×10^{23} molecules?

    c. If 0.1 mol of N_2 combine with H_2, what must be true about the quantity of H_2 for N_2 to be the limiting reactant?

_____

_____

4. _____ If a reaction's theoretical yield is 8.0 g and the actual yield is 6.0 g, what is the percent yield?

5. Joseph Priestly generated oxygen gas by strongly heating mercury(II) oxide according to the following equation:

$$2HgO(s) \xrightarrow{\Delta} 2Hg(l) + O_2(g)$$

_____ **a.** If 15.0 g HgO decompose, how many moles of HgO does this represent?

_____ **b.** How many moles of O_2 are theoretically produced?

_____ **c.** How many grams of O_2 is this?

_____ **d.** If the density of O_2 gas is 1.41 g/L, how many liters of O_2 are produced?

_____ **e.** If the percent yield is 95.0%, how many grams of O_2 are actually collected?

CHAPTER 10 REVIEW
Physical Characteristics of Gases

SECTION 10-1

SHORT ANSWER Answer the following questions in the space provided.

1. Identify whether the descriptions below describe an ideal gas or a real gas.

_____ **a.** Gas particles move in straight lines until they collide with other particles or the walls of their container.

_____ **b.** Individual gas particles have a measurable volume.

_____ **c.** The gas will not condense even when compressed or cooled.

_____ **d.** Collisions between molecules are perfectly elastic.

_____ **e.** Gas particles passing close to one another exert an attraction on each other.

2. The formula for kinetic energy is $KE = \frac{1}{2}mv^2$.

_____ **a.** What happens to the amount of kinetic energy if the mass is tripled (at constant speed)?

_____ **b.** What happens to the amount of kinetic energy if the speed is halved (at constant mass)?

_____ **c.** If two gases at the same temperature share the same kinetic energy, it follows that the molecules of greater mass have the _____ speed (faster or slower).

3. Use the kinetic-molecular theory to explain each of the following phenomena:

a. A strong-smelling gas released from a container in the middle of a room is soon detected in all areas of that room.

_____

_____

b. When 1 mol of a real gas is condensed to a liquid, the volume shrinks by a factor greater than 1000.

_____

_____

_____

SECTION 10-1 continued

c. As a gas is heated, its rate of effusion through a small hole increases if all other factors remain constant.

_____

_____

4. Explain why polar gas molecules experience larger deviations from ideal behavior than nonpolar molecules when all other factors (mass, temperature, etc) are held constant.

_____

_____

_____

_____

5. Explain the difference in the speed-distribution curves of a gas at the two temperatures shown in the figure below.

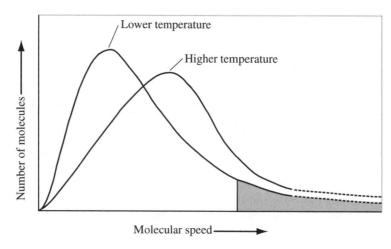

_____

_____

_____

_____

_____

CHAPTER 10 REVIEW
Physical Characteristics of Gases

SECTION 10-2

SHORT ANSWER   Answer the following questions in the space provided.

1. _____ Pressure = force/surface area. For a constant force, when the surface area is tripled the pressure _____.

   (a) is tripled
   (b) is reduced by 1/3
   (c) is unchanged

2. _____ Rank the following pressures in increasing order:

   (a) 50 kPa   **(c)** 76 torr
   (b) 2 atm   **(d)** 100 N/m^2

3. Consider the following data table:

Approximate pressure (kPa)	Altitude above sea level (km)
100	0 (sea level)
50	5.5 (peak of Mt. Kilimanjaro)
25	11 (jet cruising altitude)
< 0.1	22 (ozone layer)

   a. Explain briefly why the pressure decreases as the altitude increases.

   _____

   _____

   b. A few places on Earth are below sea level (the Dead Sea, for example). What would be true about the average atmospheric pressure there?

   _____

   _____

PROBLEMS   Write the answer on the line to the left. Show all your work in the space provided.

4. Convert a pressure of 0.200 atm to the following units:

   _____ **a.** mm Hg

_____ **b.** torr

_____ **c.** Pa

_____ **d.** kPa

5. The height of the mercury in a barometer is directly proportional to the pressure on the mercury's surface. At sea level, pressure averages 1.0 atm and the level of mercury in the barometer is 760 mm (about 30 inches). In a hurricane, the barometric reading may fall to as low as 28 inches.

_____ **a.** Convert a pressure reading of 28 inches to atmospheres.

_____ **b.** What is the barometer reading, in mm Hg, at a pressure of 0.50 atm?

c. Can a barometer be used as an altimeter (a device for measuring altitude above sea level)? Explain your answer.

_____

_____

_____

CHAPTER 10 REVIEW
Physical Characteristics of Gases

SHORT ANSWER   Answer the following questions in the space provided.

1. Explain how to correct for the partial pressure of water vapor when calculating the partial pressure of a dry gas that is collected over water.

_____

_____

_____

2. A bicycle tire is inflated to 55 lb/in.^{2} at 15°C. Assume that the volume of the tire does not change appreciably once it is inflated.

a. If the tire and the air inside it are heated to 30°C by road wear, does the pressure in the tire increase or decrease?

_____

_____

b. Because the temperature has doubled, does the pressure double to 110 psi?

_____

_____

PROBLEMS   Write the answer on the line to the left. Show all your work in the space provided.

3. _____   A 24 L sample of a gas (at fixed mass and constant temperature) exerts a pressure of 3.0 atm. What pressure will the gas exert if the volume is changed to 16 L?

4. _____   A common laboratory system to study Boyle's law uses a gas trapped in a syringe. The pressure in the system is changed by adding or removing identical weights on the plunger. The original gas volume is 50.0 mL when two weights are present. Predict the new gas volume when four more weights are added.

5. a. Use 5–6 data points from Appendix Table A-8 in the text to sketch the curve for water vapor's partial pressure versus temperature on the graph provided below.

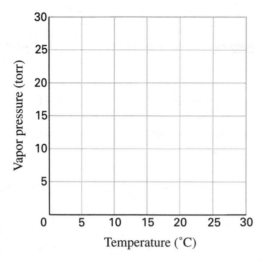

_____ **b.** Do the data points lie on a straight line?

_____ **c.** Based on your sketch, predict the approximate partial pressure for water at 11°C.

6. When an explosive like TNT is detonated, a mixture of gases at high temperature is created. Suppose that gas X has a pressure of 50 atm, gas Y has a pressure of 20 atm, and gas Z has a pressure of 10 atm.

_____ **a.** What is the total pressure in this system?

_____ **b.** Once the gas mixture combines with the air, P_{Total} soon drops to 2 atm. By what factor does the volume of the mixture increase? (Assume mass and temperature are constant.)

7. _____ A 250. mL sample of gas is collected at 57°C. What volume will the gas sample occupy at 25°C?

CHAPTER 10 REVIEW
Physical Characteristics of Gases

MIXED REVIEW

SHORT ANSWER Answer the following questions in the space provided.

1. Pressure can be represented by the following equation:

$$\text{pressure} = \frac{\text{force}}{\rule{3cm}{0.4pt}}$$

_____ **a.** For a constant area, as the force increases the pressure will _____.

_____ **b.** For a constant force, as the area increases the pressure will _____.

_____ **c.** For a constant pressure, as the area increases the force will _____.

2. Why are high-altitude research balloons only partially inflated before their launch?

_____

_____

_____

_____

_____

3. We observe that when a gas in a rigid container is warmed, the pressure on the walls increases. Explain why this occurs, based on the kinetic-molecular theory of gases.

_____

_____

_____

_____

PROBLEMS Write the answer on the line to the left. Show all your work in the space provided.

4. _____ A sample of chlorine gas is collected by water displacement at 23°C. If the atmospheric pressure is 751 torr, what is the partial pressure due to the chlorine?

MIXED REVIEW continued

5. _____ Helium gas in a balloon occupies 2.40 L at 400. K. What volume
will it occupy at 300. K?

6. _____ **a.** An air bubble with a volume of 2.0 mL forms at the bottom of a
lake, where the pressure is 3.0 atm. As it rises, the pressure on the
bubble decreases. Assume the temperature remains constant. Will
the bubble expand or contract as it rises?

_____ **b.** Predict the volume of the bubble just as it reaches the surface,
where the pressure is 1.0 atm.

7. At one point in the cycle of a piston in an automobile engine, the volume of the trapped fuel
mixture is 400. cm^3 at a pressure of 1.0 atm, and a temperature of 27°C.

_____ **a.** In the compression of the piston, the temperature reaches 77°C and
the volume decreases to 50. cm^3. What is the new pressure?

_____ **b.** After compression, a spark plug ignites the fuel mixture, starting the
power stroke of the cycle. Will the temperature of the gas go up or
down?

_____ **c.** Will the volume of the gas increase or decrease?

_____ **d.** Does the number of moles of gas remain constant throughout
the cycle?

8. _____ On a cold winter morning when the temperature is −13°C, the air
pressure in an automobile tire is 1.5 atm. If the volume does not
change, what will the pressure be after the tire has warmed to 13°C?

CHAPTER 11 REVIEW
Molecular Composition of Gases

SECTION 11-1

SHORT ANSWER   Answer the following questions in the space provided.

1. _____ The volume of a gas is directly proportional to the number of moles of that gas if _____.

    (a) the pressure remains constant
    (b) the temperature remains constant
    (c) either the pressure or temperature remains constant
    (d) both the pressure and temperature remain constant

2. _____ The molar mass of a gas at STP is the density of that gas _____.

    (a) multiplied by the mass of 1 mol   **(c)** multiplied by 22.4 L
    (b) divided by the mass of 1 mol   **(d)** divided by 22.4 L

3. _____ At constant temperature and pressure, the volumes of gaseous reactants and products can be expressed _____.

    (a) as the amount in moles of reactants minus the amount in moles of products
    (b) as ratios of small whole numbers
    (c) according to Charles's law
    (d) according to Dalton's law of partial pressures

4. Two sealed flasks of equal volume, A and B, contain two different gases at the same temperature and pressure.

_____ **a.** The two flasks must contain an equal number of molecules. True or False?

_____ **b.** The two samples must have equal masses. True or False?

    c. Now assume that flask A is warmed as flask B is cooled. Will the pressure in the two flasks remain equal? If not, which flask will have the higher pressure?

_____

_____

_____

PROBLEMS   Write the answer on the line to the left. Show all your work in the space provided.

5. _____ **a.** How many moles of methane, CH_4, are present in 5.6 L of the gas at STP?

_____ **b.** How many moles of gas are present in 5.6 L of any ideal gas at STP?

_____ **c.** What is the mass of the 5.6 L sample of CH_4?

6. _____ What is the density of CO_2 gas at STP?

7. _____ H_2 reacts according to the following equation representing the synthesis of ammonia gas:

$$N_2(g) + 3H_2(g) \rightarrow 2NH_3(g)$$

If 1 L of H_2 is consumed, what volume of ammonia will be produced at constant temperature and pressure, based on Gay-Lussac's law of combining volumes?

8. Use the data in the table below to answer the following questions:

Formula	Molar mass (g/mol)
N_2	28.02
CO	28.01
C_2H_2	26.04
He	4.00
Ar	39.95

(Assume all gases are at STP.)

_____ **a.** Which gas contains the most molecules in a 5.0 L sample?

_____ **b.** Which gas is the least dense?

_____ **c.** Which two gases have virtually the same density?

_____ **d.** What is that density measured at STP?

CHAPTER 11 REVIEW
Molecular Composition of Gases

SECTION 11-2

SHORT ANSWER Answer the following questions in the space provided.

1. _____ For the expression $V = \dfrac{nRT}{P}$, which of the following choices will cause the volume to increase?

  (a) increase P **(c)** increase T
  (b) decrease T **(d)** decrease n

2. The equation $PV = nRT$ can be rearranged to solve for any of its variables.

_____ **a.** Rearrange $PV = nRT$ to solve for P.

_____ **b.** Rearrange $PV = nRT$ to solve for T.

3. Explain how the ideal gas law can be simplified to give Avogadro's law, expressed as $\dfrac{V}{n} = k$, when the pressure and temperature of a gas are held constant.

_____

_____

_____

PROBLEMS Write the answer on the line to the left. Show all your work in the space provided.

4. _____ **a.** A large cylinder of He gas, such as that used to inflate balloons, has a volume of 25.0 L at 22°C and 5.6 atm. How many moles of He are in such a cylinder?

_____ **b.** What is the mass of the amount of helium calculated in part a?

SECTION 11-2 continued

5. ______________ **a.** Gas X has a density of 2.60 g/L at STP. Determine the molar mass of this gas.

b. Gas Y has a density of 2.60 g/L at 77°C and 0.80 atm. Its molar mass is not necessarily the same as that of gas X, even though the gases have the same density. Explain why the density is the same for X and Y but the molar masses differ.

____

____

____

______________ **c.** Determine the molar mass of gas Y.

6. ______________ **a.** How many moles of methane gas, CH_4, are present in a 4.0 g sample?

______________ **b.** Find the pressure, in torr, exerted by the CH_4 gas sample when its temperature is 27°C and its volume is 3000. mL.

7. ______________ A 7.00 L sample of argon gas at 420. K exerts a pressure of 625 kPa. If the gas is compressed to 1.25 L and the temperature is lowered to 350. K, what will be its new pressure?

CHAPTER 11 REVIEW
Molecular Composition of Gases

SECTION 11-3

SHORT ANSWER Answer the following questions in the space provided.

1. _____ Volumes of gaseous reactants and products in a chemical reaction can be expressed as ratios of small whole numbers if _____.

   (a) all reactants and products are gases
   (b) standard temperature and pressure are maintained
   (c) constant temperature and pressure are maintained
   (d) all masses represent 1 mol quantities

2. _____ In the reaction $2H_2(g) + O_2(g) \rightarrow 2H_2O(g)$, the volume ratio of H_2 to H_2O is _____.

   (a) 1:1
   (b) 2:1
   (c) 2:3
   (d) 4:3
   (e) 4:6

3. Ethanol is a component of gasohol, a type of fuel. The equation representing its complete combustion follows:

$$C_2H_6O(g) + xO_2(g) \rightarrow 2CO_2(g) + 3H_2O(l)$$

   _____ **a.** What is the value of x when the equation is correctly balanced?

   _____ **b.** Which is greater, the volume of O_2 consumed or the volume of CO_2 produced? (Assume reactants and products exist under the same conditions.)

4. The following reaction is carried out in a flexible container:

$$H_2(g) + I_2(g) \rightarrow 2HI(g)$$

The temperature and pressure are held constant throughout the reaction. Will the volume of the container increase, decrease, or remain the same as the reaction proceeds? Explain your answer.

_____

_____

_____

_____

_____

PROBLEMS Write the answer on the line to the left. Show all your work in the space provided.

5. When C_3H_4 combusts at STP, 5.6 L of C_3H_4 are consumed according to the following equation:

$$C_3H_4(g) + 4O_2(g) \rightarrow 3CO_2(g) + 2H_2O(l)$$

_____ **a.** How many moles of C_3H_4 react?

_____ **b.** How many moles of O_2, CO_2, and H_2O are either consumed or produced
_____   in the above reaction?
_____

_____ **c.** How many grams of C_3H_4 are consumed?

_____ **d.** How many liters of CO_2 are produced?

_____ **e.** How many grams of H_2O are produced?

6. _____ Chlorine in the upper atmosphere can destroy ozone molecules, O_3. The
reaction can be represented by the following equation:

$$Cl_2(g) + 2O_3(g) \rightarrow 2ClO(g) + 2O_2(g)$$

How many liters of ozone can be destroyed at 220. K and 5.0 kPa if 200.0 g
of chlorine gas react with it?

CHAPTER 11 REVIEW
Molecular Composition of Gases

SECTION 11-4

SHORT ANSWER Answer the following questions in the space provided.

1. _____ List the following gases in order of increasing rate of effusion. (Assume all gases are at the same temperature and pressure.)

      (a) He   **(b)** Xe   **(c)** HCl   **(d)** Cl_2

2. _____ The two gases in the figure below are simultaneously injected into opposite ends of the tube. They should just begin to mix closest to which labeled point?

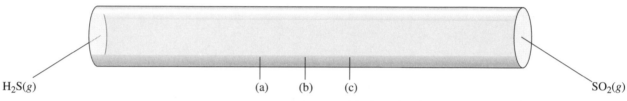

$H_2S(g)$ (a)   (b)   (c) $SO_2(g)$

3. _____ In Graham's equation, the square roots of the molar masses can be substituted with the square roots of the ____.

    (a) gas densities   **(c)** compressibilities

    (b) molar volumes   **(d)** gas constants

4. _____ The ratio of the molar masses of two gases at the same temperature and pressure can be expressed by r. What will be the value of the ratio of their average molecular speeds?

    (a) r **(d)** $\dfrac{1}{r}$

    (b) r^2 **(e)** $\left(\dfrac{1}{r}\right)^2$

    (c) $\sqrt{r}$ **(f)** $\sqrt{\dfrac{1}{}}$

5. Explain your reasoning for the order of gases you chose in item 1 above. Refer to the kinetic-molecular theory to support your explanation and cite Graham's law.

_____

_____

_____

_____

_____

_____

PROBLEMS   Write the answer on the line to the left. Show all your work in the space provided.

6. _____ A gas of unknown molar mass is observed to effuse through a small hole at one-fourth the effusion rate of hydrogen. Estimate the molar mass of this gas.

7. _____ **a.** The molar masses of He and of HCl are 4.00 g/mol and 36.46 g/mol, respectively. What is the ratio of the mass of He to the mass of HCl rounded to one decimal place?

_____ **b.** Use your answer in part a to calculate the ratio of their average speeds.

_____ **c.** If helium's average speed is 1200 m/s, what is the average speed of HCl?

8. _____ An unknown gas effuses through an opening at a rate 3.16 times slower than neon gas. Estimate the molar mass of this unknown gas.

Name _____ Date _____ Class _____

MIXED REVIEW

SHORT ANSWER Answer the following questions in the space provided.

1. _____ The average speed of a gas molecule is most directly related to the _____.

    (a) polarity of the molecule
    (b) pressure of the gas
    (c) temperature of the gas
    (d) number of moles in the sample

2. _____ Which of the following statements best represents Graham's law for two gases at the same temperature?

    (a) The molar masses and the velocities are inversely proportional.
    (b) The molar masses and the velocities are directly proportional.
    (c) The molar masses and the square roots of the velocities are directly proportional.
    (d) The molar masses and the squares of the velocities are inversely proportional.

3. Water is synthesized from its elements according to the following equation:

$$2H_2(g) + O_2(g) \rightarrow 2H_2O(l)$$

_____ **a.** Which is greater, the volume of H_2 consumed or the volume of O_2 consumed?

_____ **b.** Which is greater, the mass of H_2 consumed or the mass of O_2 consumed?

4. Consider a 2.24 L sample of an ideal gas at STP.

_____ **a.** How many moles are present in the sample, or is more data needed?

_____ **b.** How many molecules are present in the sample, or is more data needed?

_____ **c.** What is the mass of the sample, or is more data needed?

PROBLEMS Write the answer on the line to the left. Show all your work in the space provided.

5. _____ Calculate the volume of 3.5 mol of an ideal gas at 24°C and 760 torr.

6. _____ What is the molar mass of a gas that has a density of 2.68 g/L at STP?

7. A sample of oxygen gas has a mass of 8.00 g. Its boiling point is −183°C.

_____ **a.** What is the boiling point in kelvins?

_____ **b.** What is the formula and molar mass of oxygen?

_____ **c.** How many moles of oxygen are present in the 8.0 g sample?

_____ **d.** What pressure does the sample exert if it is placed in a rigid 8.21 L bottle at 77°C?

8. Hydrogen gas can be collected in the laboratory by adding an acid to aluminum metal according to the following equation:

$$2Al(s) + 6HCl(aq) \rightarrow 2AlCl_3(aq) + 3H_2(g)$$

_____ **a.** Determine the volume of H_2 that forms at 20°C and a barometric pressure of 750. torr when 9.00 g of aluminum react according to the above equation.

_____ **b.** Typically, hydrogen gas is collected by water displacement. Recalculate the volume of the dry gas taking into account the partial pressure of water vapor at the reaction temperature.

CHAPTER 12 REVIEW
Liquids and Solids

SECTION 12-1

SHORT ANSWER Answer the following questions in the space provided.

1. _____ List the following attractive forces in order of increasing strength:

      (a) hydrogen bonding   **(c)** ionic bonding

      (b) London dispersion forces   **(d)** dipole-dipole forces

2. _____ All of the following statements about liquids and gases are true *except* ____.

      (a) Molecules in a liquid are much more closely packed than molecules in a gas.

      (b) Molecules in a liquid can vibrate and rotate, but they cannot move about freely as molecules in a gas.

      (c) Liquids are much more difficult to compress into a smaller volume than are gases.

      (d) Liquids diffuse more slowly than gases.

3. _____ Liquids posses all the following properties *except* ____.

      (a) relatively low density   **(c)** relative incompressibility

      (b) the ability to diffuse   **(d)** the ability to change to a gas

4. a. Chemists distinguish between intermolecular and intramolecular forces. Explain the difference between these two types of forces.

_____

_____

Classify each of the following as intramolecular or intermolecular:

_____ **b.** hydrogen bonding in liquid water

_____ **c.** the O—H covalent bond in methanol, CH_3OH

_____ **d.** the bonds that cause gaseous Cl_2 to become a liquid when cooled

5. Explain at the molecular level the following properties of liquids:

  a. A liquid takes the shape of its container but does not expand to fill its volume.

_____

_____

_____

_____

_____

b. Polar liquids are slower to evaporate than nonpolar liquids.

_____

_____

_____

_____

_____

c. Most liquids are much denser than their corresponding gases.

_____

_____

_____

_____

_____

6. Explain briefly why liquids tend to form spherical droplets, decreasing surface area to the smallest size possible.

_____

_____

_____

_____

7. Is freezing a chemical change or a physical change? Explain your answer.

_____

_____

_____

_____

MODERN CHEMISTRY

CHAPTER 12 REVIEW
Liquids and Solids

SECTION 12-2

SHORT ANSWER Answer the following questions in the space provided.

1. Match the following descriptions on the right to the crytal type on the left.

_____ ionic crystal   **(a)** has mobile electrons in the crystal

_____ covalent molecular crystal   **(b)** is hard, brittle, and nonconducting

_____ metallic crystal   **(c)** typically has the lowest melting point of the four crystal types

_____ covalent network crystal   **(d)** has strong covalent bonds between neighboring atoms

2. For each of the four types of solids, give a specific example other than those listed in Table 12-1 on page 370 of the text.

_____

_____

_____

_____

_____

3. A chunk of solid lead is dropped into a pool of molten lead. The chunk sinks to the bottom of the pool. What does this tell you about the density of the solid lead compared with the density of the molten lead?

_____

_____

4. Answer *solid* or *liquid* to the following questions:

_____ **a.** Which is more incompressible?

_____ **b.** Which is quicker to diffuse into neighboring media?

_____ **c.** Which has a definite volume and shape?

_____ **d.** Which has molecules that are primarily rotating or vibrating in place?

5. Explain at the molecular level the following properties of solids:

   a. Solid metals conduct electricity well, but network solids do not.

   _____

   _____

   _____

   _____

   b. Almost all solids are denser than their liquid state.

   _____

   _____

   _____

   _____

   c. Amorphous solids do not have a definite melting point.

   _____

   _____

   _____

   _____

   d. Ionic crystals are much more brittle than covalent molecular crystals.

   _____

   _____

   _____

   _____

6. Experiments show that it takes 6.0 kJ of heat energy to melt 1 mol of ice at its melting point but
only about 0.6 kJ to melt 1 mol of methane, CH_4, at its melting point. Explain in terms of
intermolecular forces why it takes so much less energy to melt the methane.

   _____

   _____

   _____

   _____

   _____

CHAPTER 12 REVIEW
Liquids and Solids

SECTION 12-3

SHORT ANSWER Answer the following questions in the space provided.

1. Consider the following system at equilibrium:

$$\text{reactants} \rightleftarrows \text{products}$$

A change in conditions causes the reverse reaction to be favored.

a. What happens to the concentration of the reactants?

_____

b. What happens to the concentration of the products?

_____

2. The molar heat of vaporization of methane, CH_4, is 8.19 kJ/mol; for water, it is 40.79 kJ/mol.

_____ **a.** If 2.0×10^{23} molecules of liquid CH_4 are made to boil, how much heat must be supplied? Show all your work.

_____ **b.** Based on the molar heat of vaporization data, which is more volatile, CH_4 or H_2O?

_____ **c.** Which molecule is more polar, CH_4 or H_2O?

3. A general equilibrium equation for boiling is:

$$\text{liquid} + \text{heat energy} \rightleftarrows \text{vapor}$$

Is the forward or reverse reaction favored in each of the following cases?

_____ **a.** The temperature of the system is increased.

_____ **b.** More molecules of the vapor are added to the system.

_____ **c.** The pressure on the system is increased.

4. Consider water boiling in an open pot on a stove.

a. Can this system reach equilibrium? Why or why not?

_____

_____

b. Explain the difference between an open system and a closed system.

_____

_____

_____

5. Methanol has a normal boiling point of 65°C. It is a liquid at conditions of 1 atm and 25°C. A small beaker filled with methanol is placed under a bell jar, and the air is then pumped out. It is observed that under a vacuum the methanol boils readily at 25°C.

Use the kinetic-molecular theory and the concept of equilibrium vapor pressure to account for the lowered boiling point of methanol under a vacuum.

_____

_____

_____

_____

_____

_____

6. Refer to the phase diagram for water in Figure 12-14 on page 381 of the text to answer the following questions:

_____ **a.** Which point represents the conditions under which all three phases can coexist?

_____ **b.** Which point represents a temperature above which only the solid phase exists?

_____ **c.** Based on the diagram, as the pressure on the water system is increased, the melting point of ice ____ (increases, decreases, or stays the same).

Name _____ Date _____ Class _____

Liquids and Solids

SECTION 12-4

SHORT ANSWER Answer the following questions in the space provided.

1. Refer to the graph below to answer the following questions:

_____ **a.** What is the normal boiling point of CCl₄?

_____ **b.** What would be the boiling point of water if the air pressure over the liquid were reduced to 60 kPa?

_____ **c.** What must the air pressure over CCl₄ be for it to boil at 50°C?

d. Although water has a lower molar mass than CCl₄, it has a lower vapor pressure when measured at the same temperature. Why do you think water vapor is less volatile than CCl₄?

_____

_____

_____

_____

_____

_____

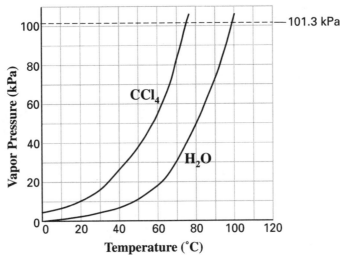

Vapor Pressure vs. Temperature for H₂O and CCl₄

2. Describe hydrogen bonding as it occurs in water in terms of the location of the bond, the particles involved, the strength of the bond, and the effects this type of bonding has on physical and chemical properties.

_____

_____

_____

_____

_____

_____

_____

PROBLEM Write the answer on the line to the left. Show all your work in the space provided.

3. A typical ice cube has a volume of about 16 cm³. Calculate the heat needed to melt the ice cube in a step by step manner. (Useful data: density of ice at 0.°C = 0.917 g/mL; heat of fusion of ice = 6.009 kJ/mol; heat of vaporization of H_2O = 40.79 kJ/mol; molar mass of H_2O = 18.02 g/mol)

_____ **a.** Determine the mass of the ice cube.

_____ **b.** Determine the number of moles of H_2O present in the sample.

_____ **c.** Determine the number of kilojoules of heat needed to melt the ice cube.

CHAPTER 12 REVIEW
Liquids and Solids

MIXED REVIEW

SHORT ANSWER   Answer the following questions in the space provided.

1. Use this general equilibrium equation to answer the following questions:

$$\text{reactants} \rightleftarrows \text{products} + \text{heat energy}$$

_____ **a.** If the reaction shifts to the right, will the concentration of reactants increase, decrease, or stay the same?

_____ **b.** If extra product is introduced, which reaction will be favored?

_____ **c.** If the temperature of the system decreases, which reaction will be favored?

2. Compare a polar water molecule with a less-polar molecule, such as formaldehyde, CH_2O. Both are liquids at room temperature and 1 atm pressure.

_____ **a.** Which liquid should have the higher boiling point?

_____ **b.** Which liquid is more volatile?

_____ **c.** Which liquid has a higher surface tension?

_____ **d.** Which liquid diffuses more rapidly?

_____ **e.** In which liquid is NaCl, an ionic crystal, likely to be more soluble?

3. Use the data table below to answer the following:

Composition	Molar mass (g/mol)	Heat of vaporization (kJ/mol)	Normal boiling point (°C)	Critical temperature (°C)
He	4	0.08	−269	−268
Ne	20	1.8	−246	−229
Ar	40	6.5	−186	−122
Xe	131	12.6	−107	+17
H_2O	18	40.8	+100	+374
HF	20	25.2	+20	+188
CH_4	16	8.9	−161	−82
C_2H_6	30	15.7	−89	+32

_____ **a.** Among nonpolar liquids, as their molar mass increases, their normal boiling point tends to ____ (increase, decrease, or stay about the same).

_____ **b.** Among compounds of approximately the same molar mass, as their polarity increases, their heat of vaporization tends to ____ (increase, decrease, or stay about the same).

c. Which is the only noble gas listed that is stable as a liquid at 0°C? Explain your answer, using the concept of critical temperature.

_____

_____

_____

PROBLEMS Write the answer on the line to the left. Show all your work in the space provided.

4. The heat of fusion of ice is 6.009 kJ/mol.

_____ **a.** How much heat is needed to melt 1.0 g of ice?

_____ **b.** An energy unit often encountered is the calorie (4.18 J = 1 calorie). Determine the heat of fusion of ice in calories/gram.

5. _____ Freon-11, CCl_3F, has been commonly used in air conditioners. Its heat of vaporization is 24.8 kJ/mol at its normal boiling point of 24°C. How much heat is removed from a room by an air conditioner that evaporates 1.00 kg of freon-11?

CHAPTER 13 REVIEW

Solutions

SECTION 13-1

SHORT ANSWER Answer the following questions in the space provided.

1. Match the type of mixture on the left to its representative particle diameter on the right.

_____ solutions   **(a)** larger than 1000 nm

_____ suspensions   **(b)** 1 nm to 1000 nm

_____ colloids   **(c)** smaller than 1 nm

2. Identify the solvent in each of the following examples:

_____ **a.** tincture of iodine (iodine dissolved in ethyl alcohol)

_____ **b.** sea water

_____ **c.** water-absorbing super gels

3. A mixture has the following properties:

- No solid settles out during a 48 hour period.
- The path of a flashlight beam is easily seen through the mixture.
- It appears to be homogeneous under a hand lens but not under a microscope.

Is the mixture a suspension, colloid, or true solution? Explain your answer.

_____

_____

_____

_____

4. Define each of the following terms:

a. alloy

_____

_____

b. electrolyte

_____

_____

    c. aerosol

_____

_____

    d. aqueous solution

_____

_____

5. For each of the following types of solutions, give an example other than those listed in Table 13-1 on page 396 of the text:

    a. a gas in a liquid

_____

_____

    b. a liquid in a liquid

_____

_____

    c. a solid in a liquid

_____

_____

6. Of the following solution models shown at the particle level, indicate which will conduct electricity. Give reasons for your answers.

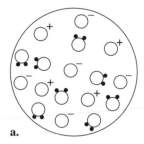 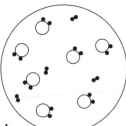 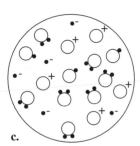

    a.   **b.**   **c.**

a. _____

b. _____

c. _____

CHAPTER 13 REVIEW

Solutions

SECTION 13-2

SHORT ANSWER   Answer the following questions in the space provided.

1. Following are statements about the dissolving process. Explain each one at the molecular level.

   a. Increasing the pressure of a solute gas above a liquid solution increases the solubility of the gas in the liquid.

_____

_____

_____

_____

   b. Increasing the temperature of water speeds up the rate at which many solids dissolve in this solvent.

_____

_____

_____

_____

_____

_____

   c. Increasing the surface area of a solid solute speeds up the rate at which it dissolves in a liquid solvent.

_____

_____

_____

_____

2. The solubility of $KClO_3$ at 25°C is 10. g of solute per 100. g of H_2O.

   a. If 15 g of $KClO_3$ are added to 100 g of water at 25°C with stirring, how much of the $KClO_3$ will dissolve? Is the solution saturated, unsaturated, or supersaturated?

_____

_____

SECTION 13-2 continued

    b. If 15 g of $KClO_3$ are added to 200 g of water at 25°C with stirring, how much of the $KClO_3$ will dissolve? Is the solution saturated, unsaturated, or supersaturated?

_____

_____

PROBLEMS Write the answer on the line to the left. Show all your work in the space provided.

3. Use the data in Table 13-4 on page 404 of the text to answer the following questions:

_____ **a.** How many grams of LiCl are needed to make a saturated solution with 300. g of water at 20°C?

_____ **b.** What is the minimum amount of water needed to dissolve 51 g of $NaNO_3$ at 40°C?

_____ **c.** Which salt forms a saturated solution when 36 g of it are dissolved in 25 g of water at 20°C?

4. KOH is an ionic solid readily soluble in water.

_____ **a.** What is its heat of solution in kJ/g? Refer to the data in Table 13-5 on page 410 of the text.

b. Will the temperature of the system increase or decrease as the dissolution of KOH proceeds? Why?

_____

_____

_____

CHAPTER 13 REVIEW

Solutions

SECTION 13-3

SHORT ANSWER Answer the following questions in the space provided.

1. Describe the errors made by the following students in making molar solutions.

   a. James needs a 0.600 M solution of KCl. He measures out 0.600 g of KCl and adds 1 L of water to the solid.

_____

_____

_____

_____

   b. Mary needs a 0.02 M solution of $NaNO_3$. She calculates that she needs 2.00 g of $NaNO_3$ for 0.02 mol. She puts this solid into a 1.00 L volumetric flask and fills the flask to the 1.00 L mark.

_____

_____

_____

_____

PROBLEMS Write the answer on the line to the left. Show all your work in the space provided.

  2. _____ What is the molarity of a solution made by dissolving 2.0 mol of solute in 6.0 L of solvent?

  3. _____ CH_3OH is soluble in water. What is the molality of a solution made by dissolving 8.0 g of CH_3OH in 250. g of water?

4. Marble chips effervesce when treated with acid. This reaction is represented by the following equation:

$$CaCO_3(s) + 2HCl(aq) \rightarrow CaCl_2(aq) + CO_2(g) + H_2O(l)$$

To produce a reaction, 25.0 mL of 4.0 M HCl is added to excess $CaCO_3$.

_____ **a.** How many moles of HCl are consumed in this reaction?

_____ **b.** How many liters of CO_2 are produced at STP?

_____ **c.** How many grams of $CaCO_3$ are consumed?

5. Tincture of iodine is $I_2(s)$ dissolved in ethanol, C_2H_5OH. A 1% solution of tincture of iodine is 10.0 g of solute for 1000. g of solution.

_____ **a.** How many grams of solvent are present in 1000. g of this solution?

_____ **b.** 10.0 g of I_2 represent how many moles of solute?

_____ **c.** What is the molality of this 1% solution?

d. To determine a solution's molarity, the density of that solution can be used. Explain how you would use the density of the tincture of iodine solution to calculate its molarity.

_____

_____

_____

_____

CHAPTER 13 REVIEW

Solutions

MIXED REVIEW

SHORT ANSWER Answer the following questions in the space provided.

1. Solid $CaCl_2$ does not conduct electricity, but it is considered to be an electrolyte. Explain.

_____

_____

_____

2. Explain the following statements at the molecular level:

a. Generally a polar liquid and a nonpolar liquid are immiscible.

_____

_____

_____

b. Carbonated soft drinks taste flat when they are warmed.

_____

_____

_____

_____

3. An unknown compound is observed to mix with toluene, $C_6H_5CH_3$, but not with water.

a. Is the unknown compound ionic, polar covalent, or nonpolar covalent?

_____

_____

b. Suppose the unknown compound is also a liquid. Will it be able to dissolve table salt? Explain your answer.

_____

_____

_____

MIXED REVIEW continued

PROBLEMS   Write the answer on the line to the left. Show all your work in the space provided.

4. Consider 500. mL of a 0.30 M $CuSO_4$ solution.

_____ **a.** How many moles of solute are present in this solution?

_____ **b.** How many grams of solute were used to prepare this solution?

5. a. If a solution is electrically neutral, can all of its ions have the same charge? Explain your answer.

_____

_____

_____ **b.** The concentration of the OH^- ions in pure water is known to be 1.0×10^{-7} M. How many OH^- ions are present in each milliliter of pure water?

6. 90. g of $CaBr_2$ are dissolved in 900. g of water.

_____ **a.** What volume does the 900. g of water occupy if its density is 1.00 g/mL?

_____ **b.** What is the molality of this solution?

CHAPTER 14 REVIEW
Ions in Aqueous Solutions and Colligative Properties

SECTION 14-1

SHORT ANSWER   Answer the following questions in the space provided.

1. Use the guidelines in Table 14-1 on page 427 of the text to predict the solubility of the following compounds in water:

_____ **a.** magnesium nitrate

_____ **b.** barium sulfate

_____ **c.** calcium carbonate

_____ **d.** ammonium phosphate

2. 1.0 mol of magnesium acetate is dissolved in water.

_____ **a.** Write the formula for magnesium acetate.

_____ **b.** How many moles of ions are released into solution?

_____ **c.** How many moles of ions are released into a solution made from 0.20 mol magnesium acetate dissolved in water?

3. In the following two precipitation reactions, write the formula for the precipitate formed:

_____ **a.** combining solutions of magnesium chloride and potassium phosphate

_____ **b.** combining solutions of sodium sulfide and silver nitrate

4. Write ionic equations showing the dissolution of the following compounds:

a. $Na_3PO_4(s)$

_____

b. iron(III) sulfate(s)

_____

5. a. Write the net ionic equation for the reaction that occurs when solutions of lead(II) nitrate and ammonium sulfate are combined.

_____

b. What are the spectator ions in this system?

_____

SECTION 14-1 continued

6. The following solutions are combined in a beaker: NaCl, Na_3PO_4, and $Ba(NO_3)_2$

   a. Write the name of each of the compounds used.

_____

_____

   b. Will a precipitate form when the above solutions are combined? If so, write the name and formula of the precipitate.

_____

_____

   c. List all spectator ions present in this system.

_____

_____

7. It is possible to have spectator ions present in many chemical systems, not just in precipitation reactions. For example:

$$Al(s) + HCl(aq) \rightarrow AlCl_3(aq) + H_2(g) \text{ (unbalanced)}$$

_____   **a.** In an aqueous solution of HCl, virtually every HCl molecule is ionized. True or False?

_____   **b.** There is only one spectator ion in this system. Is it $Al^{3+}(aq)$, $H^+(aq)$, or $Cl^-(aq)$?

   c. Balance the above equation.

_____

_____   **d.** If 9.0 g of Al metal react with excess HCl according to the balanced equation in part c, what volume of hydrogen gas at STP will be produced? Show all your work.

CHAPTER 14 REVIEW

Ions in Aqueous Solutions and Colligative Properties

SECTION 14-2

PROBLEMS Write the answer on the line to the left. Show all your work in the space provided.

1. _____ **a.** Predict the boiling point of a 0.200 m solution of glucose in water.

_____ **b.** Predict the boiling point of a 0.200 m solution of potassium iodide in water.

2. A chief ingredient of antifreeze is liquid ethylene glycol, $C_2H_4(OH)_2$. Assume $C_2H_4(OH)_2$ is added to a car radiator that holds 5.0 kg of water.

_____ **a.** How many moles of ethylene glycol should be added to the water in the radiator to lower the freezing point of that water from 0°C to −18°C?

_____ **b.** How many grams of ethylene glycol does the quantity in part a represent?

_____ **c.** Ethylene glycol has a density of 1.1 kg/L. How many liters of $C_2H_4(OH)_2$ should be added to the water in the radiator to protect the system to −18°C?

d. In World War II, soldiers in the Sahara desert needed a supply of antifreeze to protect the radiators of their vehicles. Since the temperature in the Sahara almost never drops to 0°C, why was the antifreeze necessary?

_____

_____

_____

_____

3. An important use of colligative properties is to determine the molar mass of unknown substances. The following situation is an example: 12.0 g of unknown compound X, a nonpolar nonelectrolyte, is dissolved in 100.0 g of melted camphor. The resulting solution freezes at 99.4°C. Consult Table 14-2 on page 438 of the text for any other data needed to answer the following questions:

_____ **a.** By how many Celsius degrees did the freezing point of camphor change from its normal freezing point?

_____ **b.** What is the molality of the solution of camphor and compound X based on freezing point data?

_____ **c.** If there are 12.0 g of compound X per 100.0 g of camphor, how many grams of compound X are there per kilogram of camphor?

_____ **d.** What is the molar mass of compound X?

CHAPTER 14 REVIEW

Ions in Aqueous Solutions and Colligative Properties

MIXED REVIEW

SHORT ANSWER Answer the following questions in the space provided.

1. Match the four compounds on the right to their descriptions on the left.

_____ an ionic compound that is quite soluble in water   **(a)** HCl

_____ an ionic compound that is not very soluble in water   **(b)** $BaCl_2$

_____ a molecular compound that ionizes in water   **(c)** AgCl

_____ a molecular compound that does not ionize in water   **(d)** CCl_4

2. Consider nonelectrolytes dissolved in various liquid solvents to complete the following statements:

_____ **a.** The change in the boiling point does not vary with the identity of the ____ (solute, solvent), assuming all other factors remain constant.

_____ **b.** The change in the boiling point does vary with the identity of the ____ (solute, solvent), assuming all other factors remain constant.

_____ **c.** The change in the boiling point becomes greater as the concentration of the solute in solution ____ (increases, decreases).

3. a. Name two compounds in solution that could be combined to cause calcium carbonate to precipitate.

_____

_____

b. Identify any spectator ions in the system you described in part a.

_____

c. Write the net ionic equation for the formation of calcium carbonate.

_____

4. Explain why applying rock salt (impure NaCl) to an icy sidewalk hastens the melting process.

_____

_____

_____

MIXED REVIEW continued

PROBLEMS   Write the answer on the line to the left. Show all your work in the space provided.

5. _____   Some insects survive cold winters by generating an antifreeze inside their cells. The antifreeze produced is glycerol, $C_3H_5(OH)_3$, a nonelectrolyte that is quite soluble in water. What must the molality of a glycerol solution be to lower the freezing point of water to $-25.0°C$?

6. _____   How many grams of methanol, CH_3OH, should be added to 200. g of acetic acid to lower its freezing point by $1.30°C$? Refer to Table 14-2 on page 438 of the text for any necessary data.

7. _____   The boiling point of a solution of glucose, $C_6H_{12}O_6$, and water was recorded to be $100.34°C$. Calculate the molality of this solution.

8. HF(*aq*) is a weak acid. A 0.05 mol sample of HF is added to 1.0 kg of water.

   a. Write the equation for the ionization of HF to form hydronium ions.

   _____

   _____   **b.** If HF were 100% ionized, how many moles of its ions would be released?

CHAPTER 15 REVIEW
Acids and Bases

SECTION 15-1

SHORT ANSWER Answer the following questions in the space provided.

1. Name the following compounds as acids:

_____ **a.** H_2SO_4

_____ **b.** H_2SO_3

_____ **c.** H_2S

_____ **d.** $HClO_4$

_____ **e.** hydrogen cyanide

2. _____ Which (if any) of the acids mentioned in item 1 are binary acids?

3. Write formulas for the following acids:

_____ **a.** nitrous acid

_____ **b.** hydrobromic acid

_____ **c.** phosphoric acid

_____ **d.** acetic acid

_____ **e.** hypochlorous acid

4. Calcium selenate has the formula $CaSeO_4$.

_____ **a.** What is the formula for selenic acid?

_____ **b.** What is the formula for selenous acid?

5. Use an activity series to identify two metals that will not generate hydrogen gas when treated with an acid.

_____

6. Write balanced molecular equations for the following reactions of acids and bases:

a. aluminum metal with dilute nitric acid

_____

b. calcium hydroxide solution with acetic acid

_____

SECTION 15-1 continued

7. Write net ionic equations that represent the following reactions:

a. the ionization of $HClO_3$ in water

____

b. NH_3 functioning as an Arrhenius base

____

8. a. Explain how strong acid solutions conduct an electric current.

____

____

____

b. Will a strong acid or a weak acid conduct electricity better, assuming all other factors remain constant? Explain your answer.

____

____

____

9. Most acids react with solid carbonates. For example:

$$CaCO_3(s) + HCl(aq) \rightarrow CaCl_2(aq) + H_2O(l) + CO_2(g) \text{ (unbalanced)}$$

a. Balance the above equation.

____

b. Write the net ionic equation for the above reaction.

____

______________ **c.** Identify all spectator ions in this system.

______________ **d.** How many liters of CO_2 form at STP if 5.0 g of $CaCO_3$ are treated with excess hydrochloric acid? Show all your work.

CHAPTER 15 REVIEW
Acids and Bases

SECTION 15-2

SHORT ANSWER   Answer the following questions in the space provided.

1. a. Write the two equations that show the two-stage ionization of sulfurous acid in water.

_____

_____

b. Which stage of ionization is likely favored?

_____

_____

_____

2. a. Define a Lewis base. Can OH^- function as a Lewis base? Explain your answer.

_____

_____

_____

_____

b. Define a Lewis acid. Can H^+ function as a Lewis acid? Explain your answer.

_____

_____

_____

3. a. Identify the Brønsted-Lowry acid and the Brønsted-Lowry base on the reactant side of each of the following reactions, which occur in aqueous solution. Explain your answers.

a. $H_2O(l) + HNO_3(aq) \rightarrow H_3O^+(aq) + NO_3^-(aq)$

_____

_____

b. $HF(aq) + HS^-(aq) \rightarrow H_2S(aq) + F^-(aq)$

_____

_____

4. a. Write the equation for the first ionization of H_2CO_3 in aqueous solution. Assume that water serves as the reactant that attaches to the hydrogen ion released from the H_2CO_3. Which of the reactants is the Brønsted-Lowry acid and which is the Brønsted-Lowry base? Explain your answer.

_____

_____

_____

b. Write the equation for the second ionization, that of the ion that was formed by the H_2CO_3 in the equation you just wrote. Again, assume that water serves as the reactant that attaches to the hydrogen ion released. Which of the reactants is the Brønsted-Lowry acid and which is the Brønsted-Lowry base? Explain your answer.

_____

_____

_____

c. What is the name for a substance, such as H_2CO_3, that can donate two protons?

_____

5. a. How many electron pairs surround an atom of boron (B, element 5) bonded in the compound BCl_3?

_____

b. How many electron pairs surround an atom of nitrogen (N, element 7) in the compound NF_3?

_____

c. Write an equation for the reaction between the two compounds above. Assume that they react in a 1:1 ratio to form one molecule as product.

_____

d. Assuming that the B and the N are covalently bonded to each other in the product, which of the reactants is the Lewis acid? Is this reactant also a Brønsted-Lowry acid? Explain your answers.

_____

_____

e. Which of the reactants is the Lewis base? Explain your answer.

_____

_____

Name ______________________ Date ______________ Class ______________

CHAPTER 15 REVIEW

Acids and Bases

SECTION 15-3

SHORT ANSWER **Answer the following questions in the space provided.**

1. Answer the following questions according to the Brønsted-Lowry definitions of acids and bases:

______________ **a.** What is the conjugate base of H_2SO_3?

______________ **b.** What is the conjugate base of NH_4^+?

______________ **c.** What is the conjugate base of H_2O?

______________ **d.** What is the conjugate acid of H_2O?

______________ **e.** What is the conjugate acid of $HAsO_4^{2-}$?

2. Consider the following reaction:

$$NH_4^+(aq) + CO_3^{2-}(aq) \rightleftarrows NH_3(aq) + HCO_3^-(aq)$$

a. If NH_4^+ is labeled as *acid 1*, label the other three terms as *acid 2*, *base 1*, and *base 2* to indicate the conjugate acid-base pairs.

______________ CO_3^{2-}

______________ HCO_3^-

______________ NH_3

______________ **b.** A proton has been transferred from acid 1 to base 2 in the above reaction. True or False?

3. Given the following neutralization reaction: $HCO_3^-(aq) + OH^-(aq) \rightleftarrows CO_3^{2-}(aq) + H_2O(l)$

a. Label the conjugate acid-base pairs in this system.

____

____

b. Is the forward or reverse reaction favored? Explain your answer.

____

____

____

____

SECTION 15-3 continued

4. Table 15-6 on page 471 of the text lists several amphoteric species but only one other than water is neutral.

   _____ **a.** Identify that neutral compound.

   b. Write two equations that demonstrate this compound's amphoteric properties.

   _____

   _____

   _____

5. Write the formula for the salt formed in each of the following neutralization reactions:

   _____ **a.** potassium hydroxide combined with phosphoric acid

   _____ **b.** calcium hydroxide combined with nitrous acid

   _____ **c.** hydrobromic acid combined with barium hydroxide

   _____ **d.** lithium hydroxide combined with sulfuric acid

6. Given the following unbalanced equation for a neutralization reaction:

$$H_2SO_4(aq) + NaOH(aq) \rightarrow Na_2SO_4(aq) + H_2O(l)$$

   a. Balance the equation.

   _____

   _____ **b.** In this system there are two spectator ions. Identify them.

   _____ **c.** In order to completely consume all reactants, what should be the mole ratio of acid to base?

7. The gases that produce acid rain are often referred to as NO_x and SO_x.

   a. List three specific examples of these gases.

   _____

   b. Coal- and oil-burning power plants oxidize any sulfur in their fuel as it burns in air to form SO_2 gas. The SO_2 is further oxidized by O_2 in our atmosphere to form SO_3 gas. The SO_3 gas can combine with water to form sulfuric acid. Write balanced molecular equations to illustrate these three reactions.

   _____

   _____

   _____

   c. Industrial plants making fertilizers and detergents release NO_x gases into the air. Write a balanced equation for converting $N_2O_5(g)$ into nitric acid by reacting it with water.

   _____

CHAPTER 15 REVIEW

Acids and Bases

MIXED REVIEW

SHORT ANSWER Answer the following questions in the space provided.

1. _____ **a.** Write the formula for hypochlorous acid.

_____ **b.** Write the name for HF(aq).

_____ **c.** If $Pb(C_2O_4)_2$ is lead(IV) oxalate, what is the formula for oxalic acid?

_____ **d.** Name the acid that is present in vinegar.

2. Answer the following questions according to Brønsted-Lowry acid-base theory. Consult Table 15-5 on page 468 of the text as needed.

_____ **a.** What is the conjugate base of H_2S?

_____ **b.** What is the conjugate base of HPO_4^{2-}?

_____ **c.** What is the conjugate acid of NH_3?

3. Consider the following equation:

$$OH^-(aq) + HCO_3^-(aq) \rightarrow H_2O(l) + CO_3^{2-}(aq)$$

If OH^- is base 1, label the other three terms.

_____ **a.** acid 1

_____ **b.** acid 2

_____ **c.** base 2

4. Write the formula for the salt that is produced in the following neutralization reactions:

_____ **a.** sulfurous acid combined with potassium hydroxide

_____ **b.** calcium hydroxide combined with phosphoric acid

5. Carbonic acid releases H_3O^+ ions into water in two stages.

a. Write equations representing each stage.

_____

_____

_____ **b.** Which stage releases the most ions into solution?

6. Glacial acetic acid is a highly viscous liquid that is close to 100% CH_3COOH. When it mixes with water, it forms dilute acetic acid.

   a. When making a dilute acid solution, should you add acid to water or water to acid? Explain your answer.

   _____

   _____

   _____

   b. Glacial acetic acid does not conduct electricity, but dilute acetic acid does. Explain your answer.

   _____

   _____

   _____

   c. Dilute acetic acid does not conduct electricity as well as dilute nitric acid at the same concentration. Is acetic acid a strong or weak acid?

   _____

   d. Although there are four H atoms per molecule, acetic acid is monoprotic. Show the structural formula for CH_3COOH, and indicate the H atom that ionizes.

   e. _____ How many grams of glacial acetic acid should be used to make 250 mL of 2.00 M acetic acid? Show all your work.

7. The overall effect of acid rain on lakes and ponds is partially determined by the geology of the lake bed. In some cases, the rock is limestone, which is rich in calcium carbonate. Calcium carbonate reacts with the acid in lake water according to the following (incomplete) ionic equation:

$$CaCO_3(s) + 2H_3O^+(aq) \rightarrow$$

   a. Complete the ionic equation shown above.

   _____

   b. As this reaction occurs, does the concentration of H_3O^+ in the lake water increase or decrease? What effect does this have on the acidity of the lake water?

   _____

CHAPTER 16 REVIEW

Acid-Base Titration and pH

SECTION 16-1

SHORT ANSWER Answer the following questions in the space provided.

1. Calculate the following values: (A calculator should not be necessary.)

_____ **a.** If the $[H_3O^+] = 1 \times 10^{-6}$ M for a solution, calculate the $[OH^-]$.

_____ **b.** If the $[H_3O^+] = 1 \times 10^{-9}$ M for a solution, calculate the $[OH^-]$.

_____ **c.** If the $[OH^-] = 1 \times 10^{-12}$ M for a solution, calculate the $[H_3O^+]$.

_____ **d.** If the $[OH^-]$ in part c is reduced by half, to 0.5×10^{-12} M, calculate the $[H_3O^+]$.

_____ **e.** The $[H_3O^+]$ and $[OH^-]$ are ____ (directly, inversely, or not) proportional in any system involving water.

2. Calculate the following values: (A calculator should not be necessary.)

_____ **a.** If the pH = 2.0 for a solution, calculate the pOH.

_____ **b.** If the pOH = 4.73 for a solution, calculate the pH.

_____ **c.** If the $[H_3O^+] = 1 \times 10^{-3}$ M for a solution, calculate the pH.

_____ **d.** If the pOH = 5.0 for a solution, calculate the $[OH^-]$.

_____ **e.** If the pH = 1.0 for a solution, calculate the $[OH^-]$.

3. Calculate the following values:

_____ **a.** If the $[H_3O^+] = 2.34 \times 10^{-5}$ M for a solution, calculate the pH.

_____ **b.** If the pOH = 3.5 for a solution, calculate the $[OH^-]$.

_____ **c.** If the $[H_3O^+] = 4.6 \times 10^{-8}$ M for a solution, calculate the $[OH^-]$.

PROBLEMS Write the answer on the line to the left. Show all your work in the space provided.

4. The $[H_3O^+] = 2.3 \times 10^{-3}$ M for an aqueous solution.

_____ **a.** Calculate $[OH^-]$ in this solution.

_____ **b.** Calculate the pH of this solution.

_____ **c.** Calculate the pOH of this solution.

d. Is the solution acidic, basic, or neutral? Explain your answer.

_____

5. Consider a dilute solution of 0.025 M Ba(OH)$_2$ to answer the following questions.

   a. What is the [OH$^-$] of this solution? Explain your answer.

   _____

   _____

   _____ **b.** What is the pH of this solution?

6. Vinegar purchased in a store may contain 6 g of CH$_3$COOH per 100 mL of solution.

   _____ **a.** What is the molarity of the solute?

   b. The actual [H$_3$O$^+$] of the vinegar solution in part a is 4.2×10^{-3} M. In this solution, has more than 1% or less than 1% of the acetic acid ionized? Explain your answer.

   _____

   _____

   _____ **c.** Is acetic acid strong or weak, based on the ionization information from part b?

   _____ **d.** What is the pH of this vinegar solution?

CHAPTER 16 REVIEW
Acid-Base Titration and pH

SECTION 16-2

SHORT ANSWER Answer the following questions in the space provided.

1. Below is a pH curve from an acid-base titration. On it are labeled three points: X, Y, and Z.

_____ **a.** Which point represents the equivalence point?

_____ **b.** At which point is there excess acid in the system?

_____ **c.** At which point is there excess base in the system?

_____ **d.** If the base solution is 0.250 M, how many moles of OH⁻ are consumed at the end point of the titration?

Acid-Base Titration Curve

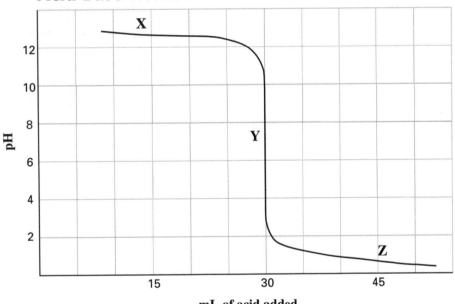

mL of acid added

PROBLEMS Write the answer on the line to the left. Show all your work in the space provided.

2. A standardized solution of 0.065 M HCl is titrated with a saturated solution of calcium hydroxide to determine its molarity and its solubility. It takes 25.0 mL of base to neutralize 10.0 mL of the acid.

a. Write the balanced molecular equation for this neutralization reaction.

_____

_____ **b.** Determine the molarity of the Ca(OH)$_2$ solution.

_____ **c.** Based on your answer to part b, calculate the solubility of the base in grams per liter of solution.

3. It is possible to carry out a titration without any indicator present. Instead, a pH probe is immersed in a beaker containing the solution of unknown molarity. The solution of known molarity is slowly added from a buret. Use the titration data below to answer the following questions:

Volume of KOH(*aq*) in the beaker = 30.0 mL

Molarity of HCl(*aq*) in the burette = 0.50 M

At the instant the pH falls from 10 to 4, 27.8 mL of acid have been added to the KOH in the beaker.

_____ **a.** What is the mole ratio of chemical equivalents in this system?

_____ **b.** Calculate the molarity of the KOH solution based on the above data.

CHAPTER 16 REVIEW
Acid-Base Titration and pH

MIXED REVIEW

SHORT ANSWER Answer the following questions in the space provided.

1. Calculate the following values: (A calculator should not be necessary.)

_____ **a.** If the $[H_3O^+] = 1 \times 10^{-4}$ M for a solution, calculate the pH.

_____ **b.** If the pH = 13.0 for a solution, calculate the $[H_3O^+]$.

_____ **c.** If the $[OH^-] = 1 \times 10^{-5}$ M for a solution, calculate the $[H_3O^+]$.

_____ **d.** If the pH = 4.72 for a solution, calculate the pOH.

_____ **e.** If the $[OH^-] = 1.0$ M for a solution, calculate the pH.

2. Calculate the following values:

_____ **a.** If the $[H_3O^+] = 6.25 \times 10^{-9}$ M for a solution, calculate the pH.

_____ **b.** If the pOH = 2.34 for a solution, calculate the $[OH^-]$.

_____ **c.** The $[OH^-]$ in milk of magnesia is ____. (Use data from Table 16-4 on page 486 of the text.)

PROBLEMS Write the answer on the line to the left. Show all your work in the space provided.

3. A 0.0012 M solution of H_2SO_4 is 100% ionized.

_____ **a.** What is the $[H_3O^+]$ of the H_2SO_4 solution?

_____ **b.** What is the $[OH^-]$ of this solution?

_____ **c.** What is the pH of this solution?

4. In a titration, a 25.0 mL sample of 0.150 M HCl is neutralized with 44.45 mL of $Ba(OH)_2$.

   a. Write the balanced molecular equation for this reaction.

   _____

   _____ **b.** What is the molarity of the base solution?

5. 3.09 g of boric acid, H_3BO_3, are dissolved into 200 mL of solution.

   _____ **a.** Calculate the molarity of the solution.

   b. H_3BO_3 ionizes in solution in three stages. Write the equation showing the ionization for each stage. Which stage proceeds furthest to completion?

   _____

   _____

   _____

   _____

   _____ **c.** What is the $[H_3O^+]$ for this boric acid solution if the pH = 4.90?

   _____ **d.** Is the percent ionization of this H_3BO_3 solution more than or less than 1%?

CHAPTER 17 REVIEW

Reaction Energy and Reaction Kinetics

SECTION 17-1

SHORT ANSWER   Answer the following questions in the space provided.

1. _____ For elements in their standard state, the value of $\triangle H_f^O$ is ____.

2. The formation and decomposition of water can be represented by the following thermochemical equations:

$$H_2(g) + \frac{1}{2}O_2(g) \rightarrow H_2O(g) + 241.8 \text{ kJ/mol}$$

$$H_2O(l) + 241.8 \text{ kJ/mol} \rightarrow H_2(g) + \frac{1}{2}O_2(g)$$

_____ **a.** Is heat being taken in or released as liquid H_2O decomposes?

_____ **b.** What is the appropriate sign for the enthalpy change in this decomposition reaction?

PROBLEMS   Write the answer on the line to the left. Show all your work in the space provided.

3. _____ If 200. g of water at 20°C absorbs 41 840 J of heat, what will its final temperature be?

4. _____ Aluminum has a specific heat of 0.900 J/(g•°C). How much energy in kJ is needed to raise the temperature of a 625 g block of aluminum from 30.7°C to 82.1°C?

5. The products in a reaction have a total heat content of 458 kJ/mol and the reactants have a total heat content of 658 kJ/mol.

_____ **a.** What is the value of $\triangle H$ for this reaction?

_____ **b.** Which is the more stable part of this system, the reactants or the products?

6. The heat of combustion of acetylene gas is -1301.1 kJ/mol of C_2H_2.

    a. Write the balanced thermochemical equation for the complete combustion of C_2H_2.

_____

_____ **b.** If 0.25 mol of C_2H_2 react according to the equation in part a, how much heat is released?

_____ **c.** How many grams of C_2H_2 are needed to react according to the equation in part a to release 3900 kJ of heat?

7. _____ When 1 mol of Al_2O_3 is formed according to the equation below, 1169.8 kJ of heat are liberated. Determine the ΔH for the reaction between Al and Fe_3O_4 if the heat of formation for Fe_3O_4 is -1120.9 kJ/mol.

$$8Al(s) + 3Fe_3O_4(s) \rightarrow 4Al_2O_3(s) + 9Fe(s)$$

8. _____ Use the data in Appendix Table A-14 of the text to determine the ΔH of the following reaction.

$$2H_2O_2(l) \rightarrow 2H_2O(l) + O_2(g)$$

CHAPTER 17 REVIEW
Reaction Energy and Reaction Kinetics

SECTION 17-2

SHORT ANSWER   Answer the following questions in the space provided.

1. For the following examples, state whether the change in entropy favors the forward or reverse reaction:

   _____ **a.** $HCl(l) \rightleftarrows HCl(g)$

   _____ **b.** $C_6H_{12}O_6(aq) \rightleftarrows C_6H_{12}O_6(s)$

   _____ **c.** $2NH_3(g) \rightleftarrows N_2(g) + 3H_2(g)$

   _____ **d.** $3C_2H_4(g) \rightleftarrows C_6H_{12}(l)$

2. _____ **a.** Write an equation that shows the relationship between enthalpy, entropy, and free energy.

   _____ **b.** For a reaction to occur spontaneously, the sign of ΔG should be ____.

3. Consider the following reaction: $NH_3(g) + H_2O(l) \rightleftarrows NH_4^+(aq) + OH^-(aq)$ + heat energy

   _____ **a.** The enthalpy factor favors the forward reaction. True or False?

   _____ **b.** The sign of $T\Delta S^o$ is negative. This means the entropy factor favors the ____.

   c. Given that ΔG^o for the above reaction is positive, which term is greater in magnitude and therefore predominates, $T\Delta S$ or ΔH?

   _____

   _____

   _____

4. Consider the following equation for the vaporization of water:

   $$H_2O(l) \rightleftarrows H_2O(g) \qquad \Delta H = +40.65 \text{ kJ/mol at } 100°C$$

   _____ **a.** Is the forward reaction exothermic or endothermic?

   _____ **b.** Does the enthalpy factor favor the forward or reverse reaction?

   _____ **c.** Does the entropy factor favor the forward or reverse reaction?

PROBLEMS Write the answer on the line to the left. Show all your work in the space provided.

5. Halogens can combine with other halogens to form several unstable compounds. Consider the following equation: $I_2(s) + Cl_2(g) \rightleftarrows 2ICl(g)$
ΔH_f° for the formation of ICl = +18.0 kJ/mol and $\Delta G^{\circ} = -5.4$ kJ/mol.

_____ **a.** Is the forward or reverse reaction favored by the enthalpy factor?

_____ **b.** Will the forward or reverse reaction occur spontaneously at standard conditions?

_____ **c.** Is the forward or reverse reaction favored by the entropy factor?

_____ **d.** Calculate the value of $T\Delta S$ for this system.

_____ **e.** Calculate the value of ΔS for this system at 25°C.

6. Calculate the free energy change for the reactions below. Determine whether each reaction will be spontaneous or nonspontaneous.

_____ **a.** $C(s) + 2H_2(g) \rightarrow CH_4(g)$

$\Delta S^O = -80.7$ J/(mol•K), $\Delta H^O = -75.0$ kJ/mol, $T = 298$ K

_____ **b.** $3Fe_2O_3(s) \rightarrow 2Fe_3O_4(s) + \frac{1}{2}O_2(g)$

$\Delta S^O = 134.2$ J/(mol•K), $\Delta H^O = 235.8$ kJ/mol, $T = 298$ K

CHAPTER 17 REVIEW

Reaction Energy and Reaction Kinetics

SECTION 17-3

SHORT ANSWER Answer the following questions in the space provided.

1. Refer to the energy diagram at the bottom of this page to answer the following questions:

_____ **a.** Which letter represents the energy of the activated complex?

    (a) A   (c) C

    (b) B   (d) D

_____ **b.** Which letter represents the energy of the reactants?

    (a) A   (c) C

    (b) B   (d) D

_____ **c.** Which of the following choices represents the quantity of activation energy for the forward reaction?

    (a) the amount of energy at C minus the amount of energy at B

    (b) the amount of energy at D minus the amount of energy at A

    (c) the amount of energy at D minus the amount of energy at B

    (d) the amount of energy at D minus the amount of energy at C

_____ **d.** Which of the following choices represents the quantity of activation energy for the reverse reaction?

    (a) the amount of energy at C minus the amount of energy at B

    (b) the amount of energy at D minus the amount of energy at A

    (c) the amount of energy at D minus the amount of energy at B

    (d) the amount of energy at D minus the amount of energy at C

_____ **e.** Which of the following choices represents the quantity of the heat of reaction for the forward reaction?

    (a) the amount of energy at C minus the amount of energy at B

    (b) the amount of energy at B minus the amount of energy at C

    (c) the amount of energy at D minus the amount of energy at B

    (d) the amount of energy at B minus the amount of energy at A

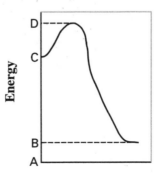

SECTION 17-3 continued

2. For the reaction $A + B \rightarrow X$, the activation energy for the forward direction equals 85 kJ/mol and the activation energy for the reverse direction equals 80 kJ/mol.

_____ **a.** Which side has the greater energy content, the reactants or the product?

_____ **b.** What is the heat of reaction in the forward direction?

_____ **c.** The heat of reaction in the reverse direction is equal in magnitude but opposite in sign to the heat of reaction in the forward direction. True or False?

3. Below is an incomplete energy diagram.

   a. Use the following data to complete the diagram: $E_a = +50$ kJ/mol; $\Delta E_{forward} = -10$ kJ/mol. Label the reactants, products, ΔE, E_a, E_a', and the activated complex.

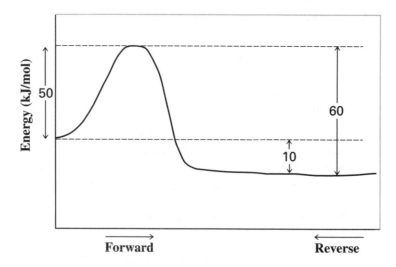

_____ **b.** What is the value of E_a'?

4. It is proposed that ozone undergoes the following two-step mechanism in our upper atmosphere:

$$O_3(g) \rightarrow O_2(g) + O(g)$$
$$O_3(g) + O(g) \rightarrow 2O_2(g)$$

a. Identify any intermediates formed in the above equations.

_____

b. Write the net equation.

_____

_____ **c.** If ΔH is negative for the reaction in part b, which is the more stable form of oxygen, O_3 or O_2?

CHAPTER **17** REVIEW

Reaction Energy and Reaction Kinetics

SECTION 17-4

SHORT ANSWER   Answer the following questions in the space provided.

1. Below is an energy diagram for a particular process. One curve represents the energy profile for the uncatalyzed reaction, and the other curve represents the energy profile for the catalyzed reaction.

_____ **a.** Which curve has the greater activation energy?

      (a) curve 1
      (b) curve 2
      (c) Both are equal.

_____ **b.** Which curve has the greater heat of reaction?

      (a) curve 1
      (b) curve 2
      (c) Both are equal.

_____ **c.** Which curve represents the catalyzed process?

      (a) curve 1
      (b) curve 2

d. Explain your answer to part c.

_____

_____

_____

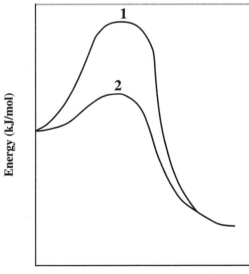

Course of reaction ⟶

SECTION 17-4 continued

2. Is it correct to say that a catalyst affects the speed of a reaction but does not take part in the reaction? Explain your answer.

_____

_____

_____

_____

_____

3. The reaction $X + Y \rightarrow Z$ is shown to have the following rate law:

$$R = k[X]^3[Y]$$

a. What is the effect on the rate if the concentration of Y is reduced by one-third?

_____

b. What is the effect on the rate if the concentration of X is doubled?

_____

c. What is the effect on the rate if a catalyst is added to the system?

_____

4. Explain the following statements using collision theory:

a. Gaseous reactants react faster under high pressure than under low pressure.

_____

_____

_____

b. Ionic compounds react faster when in solution than as solids.

_____

_____

_____

c. A class of heterogeneous catalysts called surface catalysts work best as a fine powder.

_____

_____

_____

_____

_____

MODERN CHEMISTRY

CHAPTER 17 REVIEW

Reaction Energy and Reaction Kinetics

MIXED REVIEW

SHORT ANSWER Answer the following questions in the space provided.

1. Given the following reaction for the decomposition of hydrogen peroxide:

$$2H_2O_2(l) \rightarrow 2H_2O(l) + O_2(g)$$

List three ways to speed up the rate of decomposition. For each one, briefly explain why it is effective based on collision theory.

_____

_____

_____

_____

_____

_____

2. The following equation represents a reaction that is strongly favored in the forward direction:

$$2C_7H_5(NO_2)_3(l) + 12O_2(g) \rightarrow 14CO_2(g) + 5H_2O(g) + 3N_2O(g) + \text{heat}$$

   a. Why would ΔG be negative in the above reaction?

_____

_____

_____

   b. The above reaction does not occur immediately when the reactants make contact. What does this imply about the necessary activation energy?

_____

_____

3. An ingredient in smog is the gas NO. One reaction that controls the concentration of NO in the air follows:

$$H_2(g) + 2NO(g) \rightarrow H_2O(g) + N_2O(g)$$

At high temperatures, doubling the concentration of H_2 doubles the rate of reaction, while doubling the concentration of NO increases the rate fourfold.

Write a rate law for this reaction consistent with this data.

_____

MIXED REVIEW continued

PROBLEMS   Write the answer on the line to the left. Show all your work in the space provided.

4. Consider the following equation and data: $2NO_2(g) \rightarrow N_2O_4(g)$

$$\Delta H_f^0 \text{ of } N_2O_4 = +9.2 \text{ kJ/mol}$$

$$\Delta H_f^0 \text{ of } NO_2 = +33.2 \text{ kJ/mol}$$

$$\Delta G^0 = -4.7 \text{ kJ/mol } N_2O_4$$

Use Hess's law to calculate ΔH^0 for the above reaction.

_____

_____

_____

_____

_____

5. Answer the following questions using the energy diagram at the bottom of the page.

_____ **a.** Is the reaction represented by the curve exothermic or endothermic?

_____ **b.** Estimate the magnitude and sign of $\Delta E_{forward}$.

_____ **c.** Estimate E_a'.

A catalyst is added to the reaction that lowers E_a by about 15 kJ/mol.

_____ **d.** Does the forward reaction rate speed up or slow down?

_____ **e.** Does the reverse reaction rate speed up or slow down?

_____ **f.** Does $\Delta E_{forward}$ change from its value in part b?

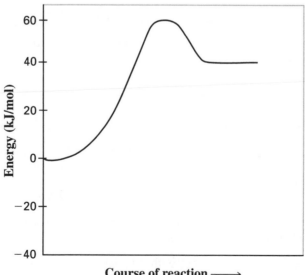

Course of reaction ⟶

CHAPTER 18 REVIEW
Chemical Equilibrium

SECTION 18-1

SHORT ANSWER Answer the following questions in the space provided.

1. _____ Silver chromate dissolves in water according to the following equation:

$$Ag_2CrO_4(s) \rightleftarrows 2Ag^+(aq) + CrO_4^{2-}(aq)$$

Which of these correctly represents the equilibrium expression for the above equation?

(a) $\dfrac{2[Ag^+] + [CrO_4^{2-}]}{Ag_2CrO_4}$ (b) $\dfrac{[Ag_2CrO_4]}{[Ag^+]^2[CrO_4^{2-}]}$ (c) $\dfrac{[Ag^+]^2[CrO_4^{2-}]}{1}$ (d) $\dfrac{[Ag^+]^2[CrO_4^{2-}]}{2[Ag_2CrO_4]}$

2. Are pure solids included in equilibrium expressions? Explain your answer.

_____

_____

_____

3. Write the equilibrium expression for the following hypothetical equation:

$$3A(aq) + B(aq) \rightleftarrows 2C(aq) + 3D(aq)$$

4. a. Write the appropriate equilibrium expression for each of the following equations. Include the value of K.

(1) $N_2O_4(g) \rightleftarrows 2NO_2(g)$ $K = 0.1$

(2) $NH_4OH(aq) \rightleftarrows NH_4^+(aq) + OH^-(aq)$ $K = 2 \times 10^{-5}$

SECTION 18-1 continued

(3) $PbI_2(s) \rightleftarrows Pb^{2+}(aq) + 2I^-(aq)$ $\qquad K = 7 \times 10^{-9}$

(4) $H_3O^+(aq) + OH^-(aq) \rightleftarrows 2H_2O(l)$ $\qquad K = 1 \times 10^{14}$

_____ **b.** Which of the four systems in part a proceeds farthest to the right when equilibrium is established?

_____ **c.** Which system contains mostly reactants at equilibrium?

PROBLEMS Write the answer on the line to the left. Show all your work in the space provided.

5. _____ Consider the following reaction:

$$2NO(g) + O_2(g) \rightleftarrows 2NO_2(g)$$

At equilibrium, $[NO] = 0.80$ M, $[O_2] = 0.50$ M, and $[NO_2] = 0.60$ M. Calculate the value of K for this reaction.

6. Nitrous acid is a weak acid that hydrolyzes according to the following equation:

$$HNO_2(aq) + H_2O(l) \rightleftarrows H_3O^+(aq) + NO_2^-(aq)$$

At 298 K, $K = 4.3 \times 10^{-4}$.

_____ **a.** Which term in the above equation does not appear in the equilibrium expression?

_____ **b.** For the above reaction, $[H_3O^+] = [NO_2^-] = 0.043$ M at equilibrium. Calculate $[HNO_2]$.

CHAPTER 18 REVIEW
Chemical Equilibrium

SECTION 18-2

SHORT ANSWER Answer the following questions in the space provided.

1. _____ Raising the temperature of any equilibrium system always results in _____.

(a) the equilibrium shifting to the left
(b) the equilibrium shifting to the right
(c) the equilibrium shifting in the direction of the exothermic reaction
(d) the equilibrium shifting in the direction of the endothermic reaction
(e) the equilibrium shifting in the direction that produces fewer molecules in the system

2. Given the equilibrium equation: $CH_3OH(g) + 101 \text{ kJ} \rightleftarrows CO(g) + 2H_2(g)$,

_____ **a.** increasing [CO] will cause [H_2] to _____.

(a) increase   (b) decrease   (c) not change

_____ **b.** raising the temperature will cause the equilibrium of the system to shift _____.

(a) to the left   (b) to the right   (c) in neither direction

_____ **c.** raising the temperature will cause the value of K to _____.

(a) increase   (b) decrease   (c) not change

3. Consider the following equilibrium equation: $H_2O(g) + C(s) \rightleftarrows H_2(g) + CO(g) + \text{heat energy}$
Which way will the equilibrium of the system shift (left, right, or in neither direction) when

_____ **a.** extra CO gas is introduced?

_____ **b.** a catalyst is introduced?

_____ **c.** the temperature of the system is lowered?

_____ **d.** the pressure on the system is increased by decreasing the container volume?

4. Adding a catalyst to a reaction does speed up the rate of that reaction. Explain why the concentration of D does not change when a catalyst is added to the reaction represented by the following general equilibrium equation: $A + B \rightleftarrows D$

_____

_____

_____

_____

5. A key step in manufacturing sulfuric acid is represented by the following equation:

$$2SO_2(g) + O_2(g) \rightleftarrows 2SO_3(g) + 100 \text{ kJ/mol}$$

To be economically viable, this process must yield as much SO_3 as possible in the shortest amount of time. You are in charge of this manufacturing process.

a. Would you impose a high pressure or a low pressure on the system? Explain your answer.

_____

_____

_____

b. To maximize the yield of SO_3, should the temperature be kept high or low during the reaction?

_____

_____

c. Will adding a catalyst change the yield of SO_3?

_____

_____

6. The equation for an equilibrium system easily studied in a lab follows:

$$2NO_2(g) \rightleftarrows N_2O_4(g)$$

N_2O_4 gas is colorless, and NO_2 gas is dark brown. Lowering the temperature of the equilibrium mixture of gases reduces the intensity of the color.

a. Is the forward or reverse reaction favored when the temperature is lowered?

_____

_____

b. Will the sign of ΔH be positive or negative if the temperature is lowered? Explain your answer.

_____

_____

_____

_____

_____

CHAPTER 18 REVIEW
Chemical Equilibrium

SECTION 18-3

SHORT ANSWER Answer the following questions in the space provided.

1. _____ The pH of tomato juice is 4.3, indicating that tomato juice is _____.

    (a) acidic
    (b) basic
    (c) neutral

2. _____ Addition of the salt of a weak acid to a solution of the weak acid _____.

    (a) lowers the concentration of the nonionized acid and the concentration of the H_3O^+
    (b) lowers the concentration of the nonionized acid and raises the concentration of the H_3O^+
    (c) raises the concentration of the nonionized acid and the concentration of the H_3O^+
    (d) raises the concentration of the nonionized acid and lowers the concentration of the H_3O^+

3. _____ Salts of a weak acid and a strong base produce solutions that are _____.

    (a) acidic only
    (b) basic only
    (c) neutral only
    (d) either acidic, basic, or neutral

4. _____ If an acid is added to a solution of a weak base and its salt, _____.

    (a) more water is formed and more weak base ionizes
    (b) hydronium-ion concentration decreases
    (c) more hydroxide ion is formed
    (d) more nonionized weak base is formed

5. a. In the space below each of the following equations, correctly label the two conjugate acid-base pairs as *acid 1, acid 2, base 1,* and *base 2.*

    (a) $CO_3^{2-}(aq) + H_3O^+(aq) \rightleftarrows HCO_3^-(aq) + H_2O(l)$

    (b) $HPO_4^{2-}(aq) + H_2O(l) \rightleftarrows OH^-(aq) + H_2PO_4^-(aq)$

    _____ **b.** Which reaction in part a is an example of hydrolysis?

    _____ **c.** As the first reaction in part a proceeds, the pH of the solution _____.

        (a) increases   (b) decreases   (c) stays at the same level

6. Write the formulas for the acid and the base that could form the salt $Ca(NO_3)_2$.

PROBLEMS   Write the answer on the line to the left. Show all your work in the space provided.

7. An unknown acid X hydrolyzes according to the equation in part a below.

  a. In the space below the equation, correctly label the two conjugate acid-base pairs in this system as *acid 1, acid 2, base 1*, and *base 2*.

$$HX(aq) + H_2O(l) \rightleftarrows X^-(aq) + H_3O^+(aq)$$

  b. Write the equilibrium expression for K_a for this system.

_____   **c.** Experiments show that at equilibrium $[H_3O^+] = [X^-] = 2.0 \times 10^{-5}$ mol/L and $[HX] = 4.0 \times 10^{-2}$ mol/L. Calculate the value of K_a based on these data.

8. Given the following equation for the reaction of a weak base in water:

$$NH_3(aq) + H_2O(l) \rightleftarrows NH_4^+(aq) + OH^-(aq)$$

Write the equilibrium expression for K_b.

CHAPTER 18 REVIEW
Chemical Equilibrium

SECTION 18-4

SHORT ANSWER   Answer the following questions in the space provided.

1. Match the solution type on the right to the corresponding relationship between the ion product and the K_{sp} for that solution listed on the left.

   _____ the ion product exceeds the K_{sp} (a) The solution is saturated; no more solid dissolves.

   _____ the ion product equals the K_{sp} (b) The solution is unsaturated; no solid is present.

   _____ the ion product is less than the K_{sp} (c) The solution is supersaturated and will readily precipitate.

2. Silver carbonate, Ag_2CO_3, makes a saturated solution with $K_{sp} = 10^{-11}$.

   a. Write the equilibrium expression for the dissolution of Ag_2CO_3.

   _____ **b.** In this system, will the foward or reverse reaction be favored if extra Ag^+ is added?

PROBLEMS   Write the answer on the line to the left. Show all your work in the space provided.

3. When the ionic solid XCl_2 dissolves in pure water to make a saturated solution, experiments show that 2×10^{-3} mol/L of X^{2+} ions go into solution.

   a. Write the equation showing the dissolution of XCl_2 and the corresponding equilibrium expression.

   _____ **b.** Calculate the value of K_{sp} for XCl_2.

_____ **c.** Refer to Table 18-3 on page 579 of the text. Would XCl_2 be more soluble or less soluble than $PbCl_2$ at the same temperature?

4. The solubility of Ag_3PO_4 is 2.1×10^{-4} g/100. g.

    a. Write the equation showing the dissolution of this ionic solid.

_____ **b.** Calculate the molarity of this saturated solution.

_____ **c.** What is the value of K_{sp} for this system?

5. As $PbCl_2$ dissolves, $[Pb^{2+}] = 2.0 \times 10^{-1}$ mol/L and $[Cl^-] = 1.5 \times 10^{-2}$ mol/L.

    a. Write the equilibrium expression for the dissolution of $PbCl_2$.

_____ **b.** Compute the ion product using the data given above.

CHAPTER 18 REVIEW
Chemical Equilibrium

MIXED REVIEW

SHORT ANSWER Answer the following questions in the space provided.

1. Consider the following equilibrium system:

$$N_2(g) + 2O_2(g) \rightleftarrows 2NO_2(g); \Delta H = +33 \text{ kJ/mol}$$

Which reaction is favored when

_____ **a.** some N_2 is removed?

_____ **b.** a catalyst is introduced?

_____ **c.** pressure on the system is increased by decreasing the volume?

_____ **d.** the temperature of the system is increased?

2. Ammonia gas dissolves in water according to the following equation:

$$NH_3(g) + H_2O(l) \rightleftarrows NH_4^+(aq) + OH^-(aq) + \text{heat}; K = 1.8 \times 10^{-5}$$

_____ **a.** Is aqueous ammonia an acid or a base?

_____ **b.** Is the equation given above an example of hydrolysis?

_____ **c.** Using the value for K does the equilibrium favor the forward or reverse reaction?

PROBLEMS Write the answer on the line to the left. Show all your work in the space provided.

3. Formic acid, HCOOH, is a weak acid present in the venom of red-ant bites. At equilibrium [HCOOH] = 2.00 M, [HCOO$^-$] = 4.0×10^{-1} M, and [H_3O^+] = $9.0 = 10^{-4}$ M.

a. Write the equilibrium expression for the ionization of formic acid.

_____ **b.** Calculate the value of K_a for this acid.

4. HF hydrolyzes according to the following equation:

$$HF(aq) + H_2O(l) \rightleftarrows H_3O^+(aq) + F^-(aq)$$

When 0.030 mol of HF dissolves in 1.0 L of water, the solution quickly ionizes to reach equilibrium. At equilibrium, the remaining [HF] = 0.027 M.

_____ **a.** How many moles of HF ionize per liter of water to reach equilibrium?

_____ **b.** What is $[F^-]$ and $[H_3O^+]$?

_____ **c.** What is the value of K_a for HF?

5. Refer to Table 18-3 on page 579 of the text. $CaSO_4(s)$ is only slightly soluble in water.

a. Write the equilibrium equation and equilibrium expression for the dissolution of $CaSO_4(s)$.

_____ **b.** Determine the solubility of $CaSO_4$ at 25°C in g/100. g H_2O.

CHAPTER 19 REVIEW

Oxidation-Reduction Reactions

SECTION 19-1

SHORT ANSWER Answer the following questions in the space provided.

1. _____ All the following equations involve redox reactions *except* _____.

    (a) $CaO + H_2O \rightarrow Ca(OH)_2$
    (b) $2SO_2 + O_2 \rightarrow 2SO_3$
    (c) $2HgO \rightarrow 2Hg + O_2$
    (d) $SnCl_4 + 2FeCl_2 \rightarrow 2FeCl_3 + SnCl_2$

2. Assign the correct oxidation number to the individual atom or ion in the following:

_____ **a.** Mn in MnO_2

_____ **b.** S in S_8

_____ **c.** Cl in $CaCl_2$

_____ **d.** I in IO_3^-

_____ **e.** C in HCO_3^-

_____ **f.** Fe in $Fe_2(SO_4)_3$

_____ **g.** S in $Fe_2(SO_4)_3$

3. In each of the following half-reactions, determine the value of x:

_____ **a.** $S^{6+} + x\,e- \rightarrow S^{2-}$

_____ **b.** $2Br^x \rightarrow Br_2 + 2e-$

_____ **c.** $Sn^{4+} + 2e- \rightarrow Sn^x$

_____ **d.** Which of the above half-reactions represent reduction processes?

4. Give examples other than those listed in Table 19-1 on page 591 of the text for the following:

_____ **a.** a compound containing H in a -1 oxidation state

_____ **b.** a peroxide

_____ **c.** a polyatomic ion where S is $+4$

_____ **d.** a substance in which F is not -1

SECTION 19-1 continued

5. OILRIG is a mnemonic device often used by students to help them understand redox reactions.

"Oxidation is loss, reduction is gain."

Explain what that phrase means—loss and gain of what?

_____

_____

6. For each of the following reactions, state whether or not any oxidation and reduction is occurring, and write the oxidation-reduction half-reactions for those cases where redox does occur:

   a. $Ca(OH)_2(aq) + 2HCl(aq) \rightarrow CaCl_2(aq) + 2H_2O(l)$

_____

_____

   b. $CH_4(g) + 2O_2(g) \rightarrow CO_2(g) + 2H_2O(g)$

_____

_____

   c. $2Al(s) + 3CuCl_2(aq) \rightarrow 2AlCl_3(aq) + 3Cu(s)$

_____

_____

7. Table 19-4 on page 615 of the text lists a half-cell reaction that represents the conversion of $Cr_2O_7^{2-}$ to Cr^{3+}.

   _____ **a.** What is the oxidation number assigned to each Cr in $Cr_2O_7^{2-}$?

   _____ **b.** How many electrons are needed to convert $2Cr^{3+}$ to $Cr_2O_7^{2-}$?

   c. Show that this half-cell reaction balances by the number of Cr, O, and H atoms present in the products and reactants.

_____

_____

CHAPTER 19 REVIEW
Oxidation-Reduction Reactions

SECTION 19-2

SHORT ANSWER Answer the following questions in the space provided.

1. _____ All of the following should be done in the process of balancing redox equations *except* ____.

    (a) adjust coefficients to balance atoms

    (b) adjust coefficients in electron equations to balance numbers of electrons lost and gained

    (c) adjust subscripts to balance atoms

    (d) write two separate electron equations

2. MnO_4^- can be reduced to MnO_2.

    _____ **a.** Assign the oxidation number to Mn in these two species.

    _____ **b.** How many electrons are gained per Mn atom in this reduction?

    _____ **c.** If 0.50 mol of MnO_4^- are reduced, how many electrons are gained?

3. Bromide ions can be oxidized to form liquid bromine. Write the balanced oxidation half-reaction for the oxidation of bromide to bromine.

_____

4. Some bleaches contain aqueous chlorine as the active ingredient. Aqueous chlorine is made by dissolving chlorine gas in water. This form of chlorine is capable of oxidizing iron(II) ions to iron(III) ions.

    a. Write the balanced ionic equation for the redox reaction between aqueous chlorine and iron(II).

_____

    b. Write the two half-reactions involved. Label them as oxidation or reduction.

_____

_____

    c. Show that the equation in part a is balanced by charge.

_____

_____

5. Balance the following equations. Write the oxidation and reduction half-reactions involved.

   a. $MnO_2(s) + HCl(aq) \rightarrow MnCl_2(aq) + Cl_2(g) + H_2O(l)$

   b. $S(s) + HNO_3(aq) \rightarrow SO_3(g) + H_2O(l) + NO_2(g)$

   c. $H_2C_2O_4(aq) + K_2CrO_4(aq) + HCl(aq) \rightarrow CrCl_3(aq) + KCl(aq) + H_2O(l) + CO_2(g)$

MODERN CHEMISTRY

CHAPTER 19 REVIEW
Oxidation-Reduction Reactions

SECTION 19-3

SHORT ANSWER Answer the following questions in the space provided.

1. For each of the following, identify the stronger oxidizing or reducing agent:

_____ **a.** Ca or Cu as a reducing agent

_____ **b.** Ag^+ or Na^+ as an oxidizing agent

_____ **c.** Fe^{3+} or Fe^{2+} as an oxidizing agent

2. For each of the following incomplete reactions, state whether a redox reaction is likely to occur:

_____ **a.** $Mg + Sn^{2+} \rightarrow$

_____ **b.** $Ag + Cu^{2+} \rightarrow$

_____ **c.** $Br_2 + I^- \rightarrow$

3. Label each of the following statements about redox as True or False:

_____ **a.** A strong oxidizing agent is itself readily reduced.

_____ **b.** In autooxidation, one chemical acts as both an oxidizing agent and a reducing agent in the same process.

_____ **c.** The number of moles of chemical oxidized must equal the number of moles of chemical reduced.

4. Solutions of Fe^{2+} are fairly unstable in part because they can undergo autooxidation, as shown by the following unbalanced equation:

$$Fe^{2+} \rightarrow Fe^{3+} + Fe$$

a. Balance the above equation.

_____

_____ **b.** If the above reaction produces 0.036 mol of Fe, how many moles of Fe^{3+} will form?

SECTION 19-3 continued

5. Oxygen gas is a powerful oxidizing agent.

   _____ **a.** Assign the oxidation number to O_2.

   _____ **b.** What does oxygen's oxidation number usually change to when it functions as an oxidizing agent?

   c. Approximately where would you place O_2 in the list of oxidizing agents in Table 19-3 on page 603 of the text?

   _____

   d. Describe the changes in oxidation states that occur in carbon and oxygen in the following combustion reaction, and identify the oxidizing and reducing agents:

$$C_6H_{12}O_6(s) + 6O_2(g) \rightarrow 6CO_2(g) + 6H_2O(l)$$

   _____

   _____

   _____

   _____

6. An example of autooxidation is the slow decomposition of aqueous chlorine, $Cl_2(aq)$, represented by the following unbalanced equation:

$$Cl_2(aq) + H_2O(l) \rightarrow ClO^-(aq) + Cl^-(aq) + H^+(aq)$$

   a. Show that the oxygen and hydrogen atoms in the above reaction are not changing oxidation states.

   _____

   _____

   b. Show the changes in the oxidation states of chlorine as this reaction proceeds.

   _____

   _____ **c.** In the oxidation reaction, how many electrons are transferred per Cl atom?

   _____ **d.** In the reduction reaction, how many electrons are transferred per Cl atom?

   e. What must be the ratio of ClO^- to Cl^- in the above reaction? Explain your answer.

   _____

   _____

   f. Balance the equation for the decomposition of $Cl_2(aq)$.

   _____

CHAPTER 19 REVIEW
Oxidation-Reduction Reactions

SECTION 19-4

SHORT ANSWER Answer the following questions in the space provided.

1. _____ In a voltaic cell, transfer of charge through the external wires occurs by means of _____.

   (a) ionization
   (b) ion migration
   (c) electron migration
   (d) proton migration

2. _____ The transfer of charge through the electrolyte solution occurs by means of _____.

   (a) ionization
   (b) ion migration
   (c) electron migration
   (d) proton migration

3. _____ All the following claims about voltaic cells are true *except* _____.

   (a) E^o_{cell} is positive.
   (b) The redox reaction in the cell occurs without the addition of electrical energy.
   (c) Electrical energy is converted to chemical energy.
   (d) Chemical energy is converted to electrical energy.

4. Label each of the following statements as applying to a voltaic cell, an electrolytic cell, or both:

   _____ **a.** The cell reaction involves oxidation and reduction.

   _____ **b.** The cell reaction proceeds spontaneously.

   _____ **c.** The cell reaction is endothermic.

   _____ **d.** The cell reaction converts chemical energy into electrical energy.

   _____ **e.** The cell reaction converts electrical energy into chemical energy.

   _____ **f.** The cell contains both a cathode and an anode.

5. Use Table 19-4 on page 615 of the text to find E^o for the following:

   _____ **a.** the reduction of MnO_4^{1-} to MnO_4^{2-}

   _____ **b.** the oxidation of Cr to Cr^{3+}

   _____ **c.** the reaction within the SHE electrode

   _____ **d.** $Cl_2 + 2Br^- \rightarrow 2Cl^- + Br_2$

SECTION 19-4 continued

PROBLEMS Write the answer on the line to the left. Show all your work in the space provided.

6. Below is a diagram of a voltaic cell.

   a. Write the anode half-reaction.

   _____

   b. Write the cathode half-reaction.

   _____

   c. Write the balanced cell reaction.

   _____

_____ **d.** Do electrons within the voltaic cell travel through the voltmeter in a clockwise or counterclockwise direction?

_____ **e.** Do anions in the beaker pass through the porous membrane in a clockwise or counterclockwise direction?

_____ **f.** Calculate what the voltmeter should read when the cell is at standard state conditions. Use data from Table 19-4 on page 615 of the text.

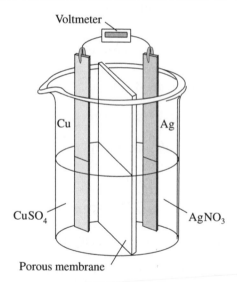

7. The silver in the voltaic cell described in item 6 is replaced with metal X and its 2+ ion. A voltage reading shows that the direction of current reverses, and the voltmeter now reads +0.74 V.

_____ **a.** From these data, calculate the reduction potential of metal X.

_____ **b.** Predict the identity of metal X, based on the data in Table 19-4 on page 615 of the text.

MODERN CHEMISTRY

CHAPTER 19 REVIEW
Oxidation-Reduction Reactions

MIXED REVIEW

SHORT ANSWER   Answer the following questions in the space provided.

1. Label the following descriptions of electrochemical cells as voltaic or electrolytic:

_____ **a.** harnesses a spontaneous redox reaction to create an electric current

_____ **b.** uses a current from an external power supply to force a nonspontaneous redox reaction to take place

_____ **c.** the reaction within an alkaline battery

_____ **d.** has a positive value for E^{o}_{cell}

PROBLEMS   Write the answer on the line to the left. Show all your work in the space provided.

2. Given the following unbalanced equation:

$$KMnO_4(aq) + HCl(aq) + Al(s) \rightarrow AlCl_3(aq) + MnCl_2(aq) + KCl(aq) + H_2O(l)$$

a. Write the oxidation and reduction half-reactions.

_____

_____

b. Balance the equation using the seven-step procedure shown in Section 19-2 of the text.

_____ **c.** Identify the oxidizing agent in this system.

_____ **d.** If this reaction were to take place in an electrochemical cell, calculate the value of E^o, using the data in Table 19-4 on page 615 of the text.

e. Is the above reaction a spontaneous redox reaction? Explain your answer.

_____

_____

3. Given the following unbalanced ionic equation: $ClO^- + H^+ \rightarrow Cl_2 + ClO_3^- + H_2O$

a. Assign the oxidation number to each term.

_____

_____

_____ **b.** How many electrons are given up by each Cl atom as they are oxidized?

_____ **c.** How many electrons are gained by each Cl atom as they are reduced?

_____ **d.** Is this an example of autooxidation?

e. Balance the above equation using the method of your choice.

CHAPTER 20 REVIEW
Carbon and Hydrocarbons

SECTION 20-1

SHORT ANSWER Answer the following questions in the space provided.

1. _____ Carbon forms four covalent bonds that are directed in space toward the corners of a regular _____.

   (a) quadrilateral   (c) polygon
   (b) pyramid   (d) tetrahedron

2. _____ The electron configuration of carbon in its ground state is _____.

   (a) $1s^2 2s^2 2p^3$ (c) $1s^2 2s^1 2p^2$
   (b) $1s^2 2s^2 2p^2$ (d) $1s^2 2s^2 2p^4$

3. _____ How many covalent bonds can a carbon atom ordinarily form?

   (a) 2   (c) 4
   (b) 3   (d) 5

4. _____ Carbon atoms form bonds readily with atoms of _____.

   (a) elements other than carbon   (c) both other elements and carbon
   (b) carbon only   (d) only neutral elements

5. _____ The bonding between atoms in a layer of graphite consists of _____.

   (a) single bonds only
   (b) double bonds only
   (c) alternating single and double bonds
   (d) bonds that are intermediate in character between single and double bonds

6. _____ Graphite is a good lubricant because it is arranged in layers that can slide across one another. The attractions that hold one layer to another are _____.

   (a) covalent attractions   (c) London dispersion attractions
   (b) ionic attractions   (d) very strong

7. _____ The hybridization of carbon's orbitals in the CH_4 molecule is _____.

   (a) sp (c) sp^3
   (b) sp^2 (d) sp^4

8. _____ The hybridization of carbon's orbitals in the C_2H_4 molecule is _____.

   (a) sp (c) sp^3
   (b) sp^2 (d) sp^4

9. _____ The hybridization of carbon's orbitals in the C_2H_2 molecule is _____.

   (a) sp (c) sp^3
   (b) sp^2 (d) sp^4

SECTION 20-1 continued

10. Explain why graphite conducts electricity, while diamond does not.

_____

_____

_____

_____

_____

11. Briefly describe the geometry of each of the allotropes of carbon.

_____

_____

_____

12. Explain why diamond conducts heat easily.

_____

_____

_____

_____

13. Identify the allotrope of carbon represented by each of the following figures:

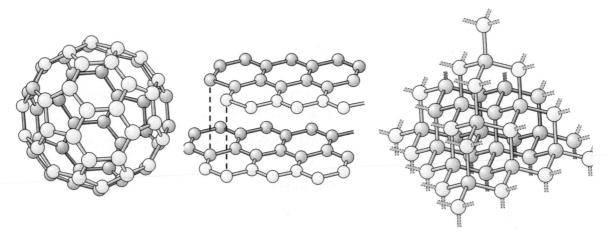

a. _____   b. _____   c. _____

CHAPTER 20 REVIEW
Carbon and Hydrocarbons

SECTION 20-2

SHORT ANSWER   Answer the following questions in the space provided.

1. Explain why the following two molecules are *not* isomers of one another.

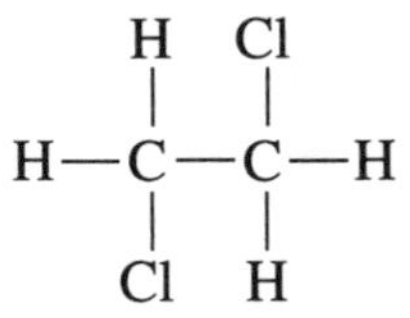

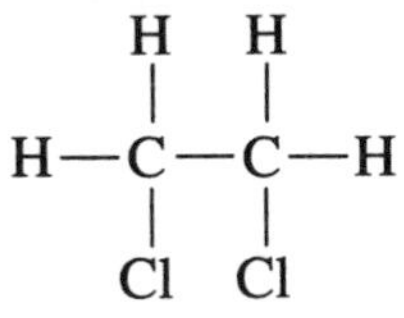

$$
\begin{array}{ccc}
\text{H} & \text{Cl} \\
| & | \\
\text{H}-\text{C}-\text{C}-\text{H} \\
| & | \\
\text{Cl} & \text{H}
\end{array}
\qquad
\begin{array}{ccc}
\text{H} & \text{H} \\
| & | \\
\text{H}-\text{C}-\text{C}-\text{H} \\
| & | \\
\text{Cl} & \text{Cl}
\end{array}
$$

_____

_____

_____

2. a. In the space below, draw the structural formula for two structural isomers with the same molecular formula.

b. In the space below, draw the structural formula for two geometric isomers with the same molecular formula.

3. Draw a structural formula that demonstrates the catenation of the methane molecule, CH_4.

4. Draw the structural formula for two structural isomers of C_4H_{10}.

5. Draw the structural formula for the *cis*-isomer of $C_2H_2Cl_2$.

6. Draw the structural formula for the *trans*-isomer of $C_2H_2Cl_2$.

CHAPTER 20 REVIEW
Carbon and Hydrocarbons

SHORT ANSWER Answer the following questions in the space provided.

1. What is a saturated hydrocarbon?

_____

_____

2. Explain why the general formula for an alkane, C_nH_{2n+2}, correctly predicts hydrocarbons in a homologous series.

_____

_____

_____

_____

3. Why is the general formula for cyclic alkanes, C_nH_{2n}, different from the general formula for straight-chain hydrocarbons?

_____

_____

4. Write the IUPAC name for the following structural formulas:

_____ **a.**

$$
\begin{array}{ccccc}
 & H & H & H & \\
 & | & | & | & \\
H - & C - & C - & C & - H \\
 & | & | & | & \\
 & H & H & H &
\end{array}
$$

_____ **b.**

$$
\begin{array}{ccccccc}
H & H & & H & & H & H \\
| & | & & | & & | & | \\
H-C-C&\rule{1cm}{0.4pt}&C&\rule{1cm}{0.4pt}&C-C-H \\
| & | & & | & & | & | \\
H & H & & H-C-H & & H & H \\
 & & & | & & & \\
 & & & H & & &
\end{array}
$$

_____ **c.**

$$
\begin{array}{cc}
CH_3-CH_2 & CH_2-CH_3 \\
| & | \\
CH_3-CH_2-CH-CH-CH_2-CH_3
\end{array}
$$

_____ **d.**

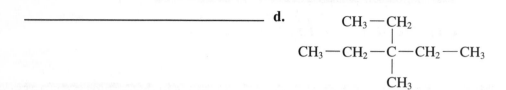

$$CH_3-CH_2$$
$$|$$
$$CH_3-CH_2-C-CH_2-CH_3$$
$$|$$
$$CH_3$$

_____ **e.** $CH_3-CH-CH_2-CH_3$
$$|$$
$$CH_3$$

5. Draw the structural formula for each of the following compounds:

a. 2,3-dimethylbutane

b. 2,2-dimethylpropane

c. 3,4-diethyl-2-methylhexane

CHAPTER 20 REVIEW
Carbon and Hydrocarbons

SECTION 20-4

SHORT ANSWER   Answer the following questions in the space provided.

1. _____ Hydrocarbons that contain double bonds are referred to as _____.

   (a) alkanes   (c) alkynes
   (b) alkenes   (d) aromatic

2. _____ Hydrocarbons that contain triple bonds are referred to as _____.

   (a) alkanes   (c) alkynes
   (b) alkenes   (d) aromatic

3. _____ Alkenes, alkynes, and aromatic hydrocarbons are all considered to be _____.

   (a) unsaturated   (c) homologous
   (b) saturated   (d) polar

4. _____ Alkenes and alkynes do not dissolve in water and are _____.

   (a) polar   (c) gases
   (b) nonpolar   (d) aromatic

5. What is an unsaturated hydrocarbon?

_____

_____

6. Why are aromatic hydrocarbons less reactive than alkenes and alkynes?

_____

_____

7. Write the IUPAC name for the following structural formulas:

_____ **a.** CH_3
                                       CH_3

_____ **b.** $CH_2-CH_2-CH_2-CH_3$
                                       CH_3

_____ **c.** 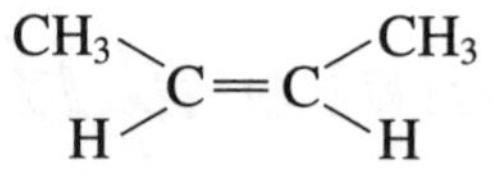 CH$_3$ \ / CH$_3$
                                        C = C
                                     H /   \ H

_____ **d.** 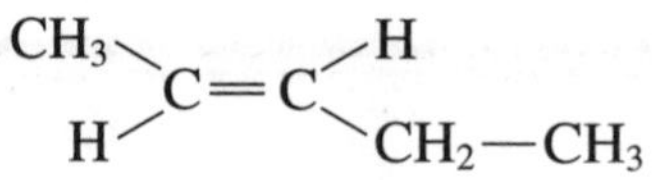 CH$_3$ \ / H
                                        C = C
                                     H /   \ CH$_2$—CH$_3$

8. Draw the structural formula for each of the following compounds:

a. 1-ethyl-2,3-dimethylbenzene

b. *trans*-3-hexene

c. *cis*-2-heptene

CHAPTER 20 REVIEW
Carbon and Hydrocarbons

MIXED REVIEW

SHORT ANSWER Answer the following questions in the space provided.

1. Identify the hybridization of carbon in each of the following molecules:

_____ **a.** C_2H_6

_____ **b.** HCN

_____ **c.** CH_2O

2. Arrange the following in order of increasing boiling point:

_____ **a.** ethane

_____ **b.** pentane

_____ **c.** 2, 2-dimethylpropane

3. Explain why hydrocarbons with only single bonds cannot form geometric isomers.

_____

_____

_____

4. How does increasing the length of a hydrocarbon chain affect the boiling point?

_____

_____

_____

_____

5. Write the IUPAC name for the following structural formulas:

a. _____

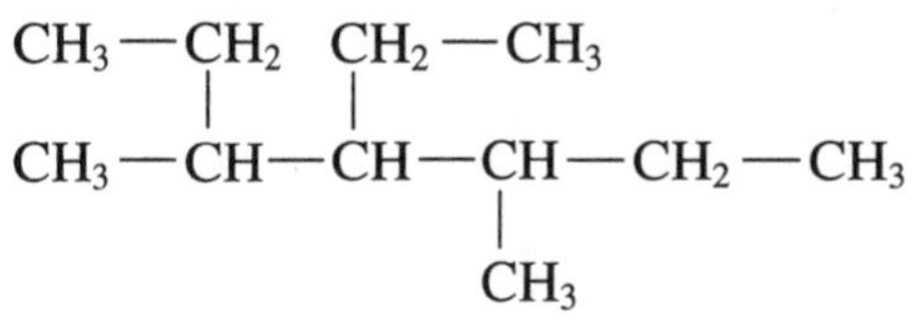

$$CH_3-CH_2 \quad CH_2-CH_3$$
$$\quad\quad\quad | \quad\quad\quad |$$
$$CH_3-CH-CH-CH-CH_2-CH_3$$
$$\quad\quad\quad\quad\quad\quad\quad |$$
$$\quad\quad\quad\quad\quad\quad\quad CH_3$$

b. _____

$$CH_2{=}CH-CH_2-CH_2-CH_2-CH_3$$

c. _____
$$CH_3 - CH - CH = CH_2$$
with CH_3 attached above the second carbon:

$$\begin{array}{c} CH_3 \\ | \\ CH_3-CH-CH=CH_2 \end{array}$$

d. _____ $CH_3 - C \equiv C - CH_3$

6. Draw the structural formula for each of the following compounds:

a. 1-ethyl-3-methylbenzene

b. 1,2,4-trimethylcyclohexane

c. 3-methyl-1-pentyne

d. 3-ethyl-2-methylhexane

CHAPTER **21** REVIEW
Other Organic Compounds

SECTION 21-1

SHORT ANSWER Answer the following questions in the space provided.

1. Draw the structural formula for each of the following compounds:

   a. 1-butanol

   b. 1,1,2-trifluoroethene

   c. dipropyl ether

2. Write the IUPAC name for the following structural formulas:

_____ **a.**

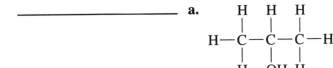

_____ **b.**

$$H - \underset{\underset{\displaystyle OH}{|}}{\overset{\overset{\displaystyle H}{|}}{C}} - \underset{\underset{\displaystyle OH}{|}}{\overset{\overset{\displaystyle H}{|}}{C}} - \underset{\underset{\displaystyle H}{|}}{\overset{\overset{\displaystyle H}{|}}{C}} - H$$

3. There are four isomers in the alcohol family that fit the formula C_4H_9OH. Draw the structural formula for each isomer, and give each its corresponding IUPAC name.

PROBLEM **Write the answer on the line to the left. Show all your work in the space provided.**

4. MTBE, an important gasoline additive, is discussed in Section 21-1 of the text. Its formula is $(CH_3)_3COCH_3$.

_____ **a.** What is its molar mass?

_____ **b.** What is the mass of 0.25 mol of MTBE?

CHAPTER 21 REVIEW
Other Organic Compounds

SECTION 21-2

SHORT ANSWER Answer the following questions in the space provided.

1. Match the structural formula on the right to the corresponding family name on the left.

_____ aldehyde

_____ ketone

_____ carboxylic acid

_____ amine

_____ ester

_____ alkene

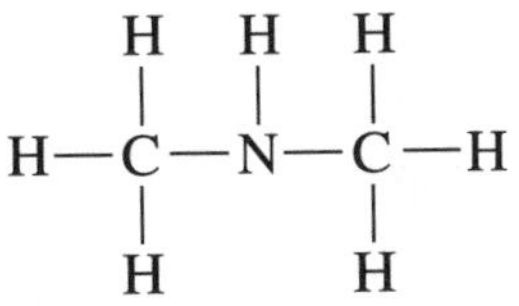

(a) H—C—OH
        ‖
        O

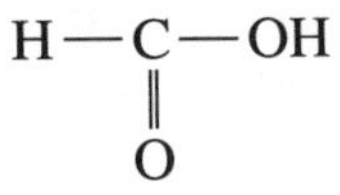

(b)   H   H   H
          | | |
     H—C—N—C—H
          | |
          H   H

(c)   H
                  |
     H—C—O—C—H
        ‖ |
        O   H

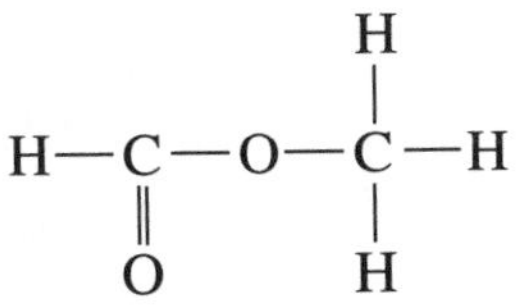

(d)   H
           |
     H—C—C=O
           | |
           H   H

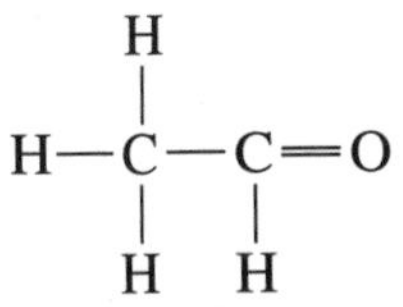

(e) H\   /H
       C=C
    H/   \H

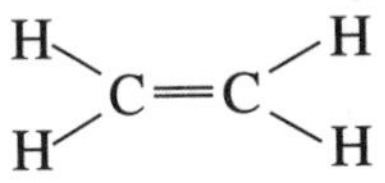

(f)   H   O   H
       | ‖ |
  H—C—C—C—H
       | |
       H   H

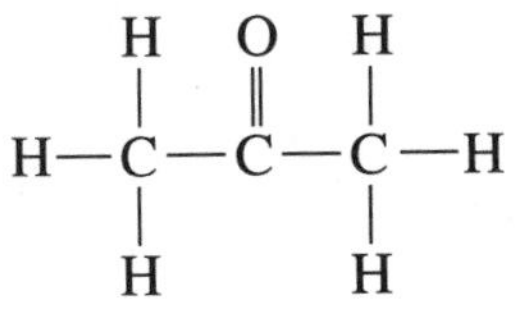

2. Draw the structural formula for each of the following compounds:

a. 2-pentanone

b. 3-methyl hexanoic acid

    c. propyl amine

    d. methyl butanoate

3. Write the IUPAC name for the following structural formulas:

_____ **a.**

$$H-\overset{\overset{\displaystyle O}{\|}}{C}-\overset{\overset{\displaystyle H}{|}}{\underset{\underset{\displaystyle H}{|}}{C}}-\overset{\overset{\displaystyle H}{|}}{\underset{\underset{\displaystyle H}{|}}{C}}-H$$

_____ **b.**

$$H-\overset{\overset{\displaystyle H}{|}}{\underset{\underset{\displaystyle H}{|}}{C}}-\overset{\overset{\displaystyle O}{\|}}{C}-OH$$

_____ **c.**

$$H-\overset{\overset{\displaystyle H}{|}}{\underset{\underset{\displaystyle H}{|}}{C}}-\overset{\overset{\displaystyle O}{\|}}{C}-\overset{\overset{\displaystyle Br}{|}}{\underset{\underset{\displaystyle H}{|}}{C}}-\overset{\overset{\displaystyle H}{|}}{\underset{\underset{\displaystyle H}{|}}{C}}-\overset{\overset{\displaystyle H}{|}}{\underset{\underset{\displaystyle H}{|}}{C}}-H$$

_____ **d.**

$$H-\overset{\overset{\displaystyle H}{|}}{\underset{\underset{\displaystyle H}{|}}{C}}-\overset{\overset{\displaystyle H}{|}}{\underset{\underset{\displaystyle H-C-H}{|}}{C}}-\overset{\overset{\displaystyle H}{|}}{\underset{\underset{\displaystyle H}{|}}{C}}-\overset{\overset{\displaystyle H}{|}}{N}-H$$

MODERN CHEMISTRY

CHAPTER 21 REVIEW
Other Organic Compounds

SECTION 21-3

SHORT ANSWER   Answer the following questions in the space provided.

1. _____ A saturated organic compound _____.

    (a) contains all single bonds
    (b) contains at least one double or triple bond
    (c) contains only carbon and hydrogen atoms
    (d) is quite soluble in water

2. Match the reaction type on the left to its description on the right.

_____ substitution   **(a)** An atom or molecule is added to an unsaturated molecule, increasing the saturation of the molecule.

_____ addition   **(b)** A simple molecule is removed from adjacent carbon atoms of a larger molecule.

_____ condensation   **(c)** One or more atoms replace another atom or group of atoms in a molecule.

_____ elimination   **(d)** Two molecules or parts of the same molecule combine.

3. Substitution reactions can require a catalyst to be feasible. The reaction represented by the following equation is heated to maximize the percent yield:

$$C_2H_6(g) + Cl_2(g) \overset{\Delta}{\rightleftarrows} C_2H_5Cl(l) + HCl(g)$$

_____ **a.** Should a high or low temperature be maintained?

_____ **b.** Should a high or low pressure be used?

_____ **c.** Should the HCl gas be allowed to escape?

4. Elemental bromine is a red-brown liquid that is colorless when it exists in compounds. A qualitative test for carbon-carbon multiple bonds is to add a few drops of bromine solution to a hydrocarbon sample at room termperature and in the absence of sunlight. The bromine will either quickly lose its color or remain reddish brown.

_____ **a.** If the sample is unsaturated, what type of reaction should occur when the bromine is added under the conditions mentioned above?

_____ **b.** If the sample is saturated, what type of reaction should occur when the bromine is added under the conditions mentioned above?

_____ **c.** The red-brown color of a bromine solution added to a hydrocarbon sample at room temperature and in the absence of light quickly disappears. Is the sample a saturated or unsaturated hydrocarbon?

5. Write the formulas for the products of the following reactions:

   a. addition of chlorine to ethene

_____

   b. substitution of iodine with ethane

_____

   c. elimination of water from

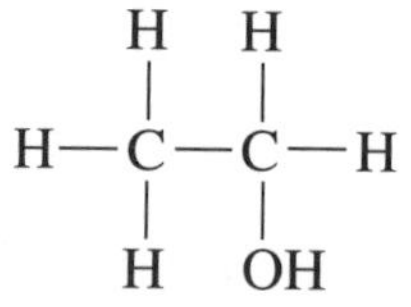

$$H-\overset{\overset{\displaystyle H}{|}}{\underset{\underset{\displaystyle H}{|}}{C}}-\overset{\overset{\displaystyle H}{|}}{\underset{\underset{\displaystyle OH}{|}}{C}}-H$$

_____

6. Two glucose molecules, $C_6H_{12}O_6$, undergo a condensation reaction to form one molecule of sucrose, $C_{12}H_{22}O_{11}$.

   _____ **a.** How many molecules of water split off during this condensation reaction?

   b. Write a balanced formula equation for this condensation reaction.

_____

7. Addition reactions with halogens tend to proceed rapidly and easily with the two halogen atoms bonding to the carbon atoms connected by the multiple bond. Thus, only one isomeric product forms.

   a. Write an equation showing the structural formulas for the reaction of Br_2 with 1-butene.

   b. Name the product.

_____

   _____ **c.** What is the mole ratio of 1-butene to Br_2 in the above reaction?

   d. How many moles of Br_2 will add to 1 mol of 1-butyne? Explain your answer.

_____

_____

_____

CHAPTER 21 REVIEW
Other Organic Compounds

SECTION 21-4

SHORT ANSWER Answer the following questions in the space provided.

1. Identify each of the following substances as either a natural or a synthetic polymer:

_____ **a.** cellulose

_____ **b.** nylon

_____ **c.** proteins

_____ **d.** polyisoprene

_____ **e.** kevlar

_____ **f.** polypropylene

2. Table 21-8 on page 687 of the text shows the styrene monomer, $CH_2{=}CH{-}R$, where $R = $ ⬡.

_____ **a.** What is the formula of R in this molecule?

_____ **b.** What is the molar mass of this styrene monomer?

3. The text gives several abbreviations commonly used in describing plastics or polymers. For each of the following abbreviations, give the complete name and one common usage:

a. HDPE

_____

b. LDPE

_____

c. CLPE

_____

d. PVA

_____

e. PVC

_____

f. SBR

_____

SECTION 21-4 continued

g. PET

____

4. Sulfur is often used in making cross-linked polymers (vulcanized rubber, for example).

a. Give the orbital electron arrangement of a neutral sulfur atom.

____

______ **b.** How many unpaired electrons are there in this atom?

c. Explain how sulfur links two adjacent chains of polyethylene.

____

____

____

5. Based on the description of natural rubber in the text, is natural rubber a thermoplastic or thermosetting polymer? Explain your answer.

____

____

6. Explain why an alkane cannot be used as the monomer of an addition polymer.

____

____

____

CHAPTER 21 REVIEW

Other Organic Compounds

MIXED REVIEW

SHORT ANSWER Answer the following questions in the space provided.

1. Match the general formula on the right to the corresponding family name on the left.

_____ carboxylic acid

_____ ester

_____ alcohol

_____ ether

_____ alkyl halide

_____ amine

_____ aldehyde

_____ ketone

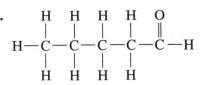

(a) $R\!-\!X$

(b) $R\!-\!\overset{\displaystyle O}{\overset{\|}{C}}\!-\!H$

(c) $R\!-\!O\!-\!R'$

(d) $R\!-\!OH$

(e) $R\!-\!\overset{\displaystyle O}{\overset{\|}{C}}\!-\!OH$

(f) $R\!-\!\overset{\displaystyle\;}{\underset{\displaystyle R'}{N}}\!-\!R''$

(g) $R\!-\!\overset{\displaystyle O}{\overset{\|}{C}}\!-\!R'$

(h) $R\!-\!\overset{\displaystyle O}{\overset{\|}{C}}\!-\!O\!-\!R'$

2. Write the IUPAC name for the following structural formulas:

_____ **a.**

$$H-\overset{\displaystyle H}{\underset{\displaystyle H}{C}}-\overset{\displaystyle H}{\underset{\displaystyle H}{C}}-\overset{\displaystyle H}{\underset{\displaystyle H}{C}}-\overset{\displaystyle H}{\underset{\displaystyle H}{C}}-\overset{\displaystyle O}{\overset{\|}{C}}-H$$

_____ **b.**

$$H-\overset{\displaystyle H}{\underset{\displaystyle H}{C}}-\overset{\displaystyle H}{\underset{\displaystyle H}{C}}-\overset{\displaystyle \;}{\underset{\displaystyle O}{\overset{\|}{C}}}-OH$$

_____ **c.**

$$H-\overset{\displaystyle Cl}{\underset{\displaystyle H}{C}}-\overset{\displaystyle Cl}{\underset{\displaystyle H}{C}}-\overset{\displaystyle H}{\underset{\displaystyle H}{C}}-\overset{\displaystyle Cl}{\underset{\displaystyle H}{C}}-H$$

_____ **d.**

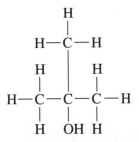

MIXED REVIEW continued

3. Write the name of the organic product formed by the following reactions:

______________________ **a.** addition of bromine to 2-butene

______________________ **b.** elimination of water from 1-propanol

______________________ **c.** reaction of methane and F_2 in direct sunlight

4. Recall that isomers in organic chemistry have identical molecular formulas but different structures and IUPAC names.

______________ **a.** Two isomers must have the same molar mass. True or False?

______________ **b.** Two isomers must have the same boiling point. True or False?

______________ **c.** Are ethanol and dimethyl ether isomers? Support your answer by drawing the structural formula for each compound and labeling it.

5. Each of the following names implies a structure but is not a correct IUPAC name. For each example, draw the implied structural formula and write the correct IUPAC name.

a. 3-bromopropane

b. ethylethylether

c. 4-butanol

CHAPTER 22 REVIEW

Nuclear Chemistry

SECTION 22-1

SHORT ANSWER   Answer the following questions in the space provided.

1. _____ Based on the masses of the three elementary particles reported in Section 22-1 on page 701 of the text, which has the greatest mass?

    (a) the proton
    (b) the neutron
    (c) the electron

2. _____ The force that keeps nucleons together is _____.

    (a) a strong force
    (b) a weak force
    (c) an electromagnetic force
    (d) gravity

3. _____ The stability of a nucleus is most affected by the _____.

    (a) number of neutrons
    (b) number of protons
    (c) number of electrons
    (d) ratio of neutrons to protons

4. _____ If an atom should form from its constituent particles, _____.

    (a) matter is lost and energy is taken in
    (b) matter is lost and energy is released
    (c) matter is gained and energy is taken in
    (d) matter is gained and energy is released

5. _____ For atoms of a given mass number, the greater the mass defect, the _____.

    (a) smaller the binding energy per nucleon
    (b) greater the binding energy per nucleon
    (c) binding energy per nucleon stays the same

6. Based on the data in Table 22-1 on page 705 of the text, which isotope of He, helium-3 or helium-4,

_____ **a.** has the smaller binding energy per nucleon?

_____ **b.** is more stable to nuclear changes?

7. _____ The number of neutrons in an atom of magnesium-25 is _____.

8. _____ The mass number of a given nuclide is A, and its atomic number is Z. Use these two variables to write the formula to calculate the number of neutrons in a nuclide.

SECTION 22-1 continued

9. Atom X has 50 nucleons and a binding energy of 4.2×10^{-11} J. Atom Z has 80 nucleons and a binding energy of 8.4×10^{-11} J.

_____ **a.** The mass defect of Z is twice that of X. True or False?

_____ **b.** Which atom has the greater binding energy per nucleon?

_____ **c.** Which atom is likely to be more stable to nuclear transmutations?

10. Identify the missing term in the following nuclear equations. Write the element's symbol, its atomic number, and its mass number.

_____ **a.** $^{14}_{6}C \rightarrow {}^{0}_{-1}e +$ _____

_____ **b.** $^{63}_{29}Cu + {}^{1}_{1}H \rightarrow$ _____ $+ {}^{4}_{2}He$

11. Write the equation that shows the equivalency of mass and energy.

_____

12. Consider the two nuclides $^{56}_{26}Fe$ and $^{14}_{6}C$.

   a. Determine the number of protons in each nuclei.

   _____

   b. Determine the number of neutrons in each nuclei.

   _____

   c. Determine whether the $^{56}_{26}Fe$ nuclide is likely to be stable or unstable, based on its position in the band of stability shown in Figure 22-2 on page 703 of the text.

   _____

PROBLEM Write the answer on the line to the left. Show all your work in the space provided.

13. _____ Neon-20 is a stable isotope of neon. Its actual mass has been found to be 19.992 44 amu. Determine the mass defect in this nuclide.

CHAPTER 22 REVIEW
Nuclear Chemistry

SECTION 22-2

SHORT ANSWER Answer the following questions in the space provided.

1. _____ The nuclear equation $^{210}_{84}Po \rightarrow \, ^{206}_{82}Pb + \, ^{4}_{2}He$ is an example of ____.

    (a) alpha emission
    (b) beta emission
    (c) positron emission
    (d) electron capture

2. _____ $^{a}_{b}Z$ undergoes electron capture to form a new element X. Which of the following choices best represents the product formed?

    (a) $^{a-1}_{b}X$
    (b) $^{a+1}_{b}X$
    (c) $^{a}_{b+1}X$
    (d) $^{a}_{b-1}X$

3. _____ Which of the following choices best represents the fraction of a radioactive sample remaining after four half-lives have occurred?

    (a) $(1/2)^4$
    (b) $(1/2) \times 4$
    (c) $(1/4)$
    (d) $(1/4)^2$

4. Match the nuclear symbol on the right to the name of the corresponding particle on the left.

    _____ beta particle   **(a)** $^{1}_{1}H$

    _____ proton   **(b)** $^{4}_{2}He$

    _____ positron   **(c)** $^{0}_{-1}\beta$

    _____ alpha particle   **(d)** $^{0}_{+1}\beta$

5. Label each of the following statements as True or False. Each member of a decay series ____.

    _____ **a.** shares the same atomic number

    _____ **b.** differs in mass number from others by multiples of 4

    _____ **c.** has a unique atomic number

    _____ **d.** differs in atomic number from others by multiples of 4

Name _____ Date _____ Class _____

6. _____ Identify the missing term in the following nuclear equation. Write the element's symbol, its atomic number, and its mass number.

$$\underline{\quad} \rightarrow {}^{231}_{90}\text{Th} + {}^{4}_{2}\text{He}$$

7. Lead-210 undergoes beta emission. Write the nuclear equation showing this transmutation.

_____

8. Einsteinium is a transuranium element. Einsteinium-247 can be prepared by bombarding uranium-238 with nitrogen-14 nuclei, releasing several neutrons, as shown by the following equation:

$$^{238}_{92}\text{U} + {}^{14}_{7}\text{N} \rightarrow {}^{247}_{99}\text{Es} + x \, {}^{1}_{0}n$$

What must be the value of x in the above equation? Explain your reasoning.

_____

_____

_____

PROBLEMS   Write the answer on the line to the left. Show all your work in the space provided.

9. _____ Phosphorus-32 has a half-life of 14.3 days. How many days will it take for a radioactive sample to decay to one-eighth its original size?

10. _____ Iodine-131 has a half-life of 8.0 days. How many grams of an original 160 mg sample will remain in 40 days?

11. _____ Carbon-14 has a half-life of 5715 years. It is used to determine the age of ancient objects. If a sample today contains 0.060 mg of carbon-14, how much carbon-14 must have been present in the sample 11 430 years ago?

CHAPTER **22** REVIEW

Nuclear Chemistry

SECTION **22-3**

SHORT ANSWER Answer the following questions in the space provided.

1. _____ The radioisotope cobalt-60 is used for all of the following applications *except* to ____.

    (a) kill food-spoiling bacteria   **(d)** treat heart disease
    (b) preserve food   **(e)** treat certain kinds of cancers
    (c) kill insects that infest food

2. _____ All of the following contribute to background-radiation exposure *except* ____.

    (a) radon in homes and buildings
    (b) cosmic rays passing through the atmosphere
    (c) consumption of irradiated foods
    (d) X rays obtained for medical or dental reasons
    (e) rocks in Earth's soil

3. _____ Review the Dating Game Commentary on page 716 of the text. Which of the graphs shown below best illustrates the decay of a sample of carbon-14? Assume each division on the time axis represents 5715 years.

(a)
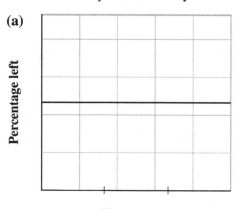

(c)
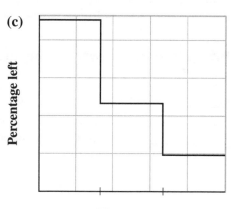

(b)
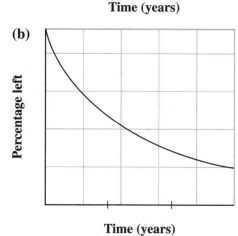

(d)
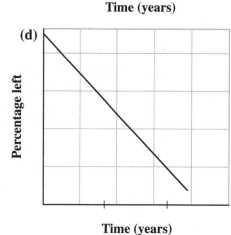

SECTION 22-3 continued

4. Match the item on the left with its description on the right.

_____ Geiger-Müller counter

_____ scintillation counter

_____ film badge

_____ radioactive tracer

(a) uses exposure to film to measure the approximate radiation exposure of people working with radiation

(b) instrument that converts scintillating light to an electric signal for detecting radiation

(c) detects radiation by counting electric pulses carried by gas ionized by radiation

(d) radioactive atoms that are incorporated into substances so that movement of the substances can be followed by detectors

5. Which type of radiation is most easily absorbed by shielding? Why?

_____

_____

6. One technique for dating ancient rocks involves uranium-235, which has a half-life of 710 million years. Rocks originally rich in uranium-235 will contain small amounts of its decay series, including the nonradioactive lead-206. Explain the relationship between the ratio of lead-206 to uranium-235 in a sample and the sample's age.

_____

_____

PROBLEMS Write the answer on the line to the left. Show all your work in the space provided.

7. _____ The technetium-99 isotope, described in Figure 22-13 on page 715 of the text, has a half-life of 6.0 h. If a 100. mg sample of technetium-99 were injected into a patient, how many milligrams would still be present after one day?

8. _____ A Geiger-Müller counter used to detect carbon-14 atoms registers 14 units when exposed to a living organism. What would the counter read in units if that same organism had been dead for 11 430 years? (Hint: The half-life of carbon-14 is 5715 years.)

CHAPTER 22 REVIEW
Nuclear Chemistry

SECTION 22-4

SHORT ANSWER Answer the following questions in the space provided.

1. Label each of the following statements with one of the choices below:

(1) fission only   (3) both fission and fusion
(2) fusion only   (4) neither fission nor fusion

_____ **a.** A very large nucleus splits into smaller pieces.

_____ **b.** The total mass before a reaction is greater than the mass after a reaction.

_____ **c.** The rate of a reaction can be safely controlled in suitable vessels.

_____ **d.** Two small nuclei form a single larger one.

_____ **e.** Less-stable nuclei are converted to more-stable nuclei.

2. Match the reaction type on the right to the statement that applies to it on the left.

_____ requires very high temperatures   **(a)** uncontrolled fusion

_____ used in nuclear reactors to make electricity   **(b)** uncontrolled fission

_____ used in hydrogen bombs   **(c)** controlled fusion

_____ used in atomic bombs   **(d)** controlled fission

3. Match the component of a nuclear power plant on the right to its use on the left.

_____ limits the number of free neutrons   **(a)** moderator

_____ used to slow down neutrons   **(b)** fuel rod

_____ drives an electrical generator   **(c)** control rod

_____ provides neutrons by its fission   **(d)** shielding

_____ removes heat from the system safely   **(e)** coolant

_____ prevents escape of radiation   **(f)** turbine

4. _____ A chain reaction is any reaction in which _____.

     (a) excess reactant is present   **(c)** the rate is slow

     (b) the material that starts the reaction is also a product   **(d)** many steps are involved

5. As a star ages, does the ratio of He atoms to H atoms in its composition become larger, smaller, or remain constant? Explain your answer.

_____

_____

_____

_____

6. The products of nuclear fission are variable; many possible nuclides can be created. In the feature An Unexpected Finding, on page 720 of the text, it was noted that Meitner showed radioactive barium to be one product of fission. Following is an incomplete possible nuclear equation to produce barium-141:

$$^{235}_{92}U + ^{1}_{0}n \rightarrow ^{141}_{56}Ba + \underline{\quad\quad} + 4\,^{1}_{0}n + \text{energy}$$

_____ **a.** Determine the missing fission product formed. Write the element's symbol, its atomic number, and its mass number.

_____ **b.** Is it likely that this isotope in part a is unstable? (Refer to Figure 22-2 on page 703 of the text.)

7. Small nuclides other than hydrogen-1 can undergo fusion.

_____ **a.** Complete the following nuclear equation by identifying the missing term. Write the element's symbol, its atomic number, and its mass number.

$$^{3}_{1}H + ^{7}_{3}Li \rightarrow \underline{\quad\quad} + \text{energy}$$

b. When measured exactly, the total mass of the reactants does not add up to that of the products in the reaction represented in part a. Why is there a difference between the mass of the products and the mass of the reactants? Which has the greater mass, the reactants or the products?

_____

_____

_____

8. What are some current concerns regarding nuclear-power-plant development?

_____

_____

_____

_____

CHAPTER 22 REVIEW
Nuclear Chemistry

MIXED REVIEW

SHORT ANSWER Answer the following questions in the space provided.

1. The ancient alchemists dreamed of being able to turn lead into gold. By using lead-206 as the target atom of a powerful accelerator, we can attain that dream in principle. Find a one-step process that will convert $^{206}_{82}Pb$ into a nuclide of gold-79. You may use alpha particles, beta particles, positrons, or protons. Write the nuclear equation to turn lead into gold.

_____

2. A typical fission reaction releases 2×10^{10} kJ/mol of uranium-235, while a typical fusion reaction produces 6×10^8 kJ/mol of hydrogen-1. Which process produces more energy per gram of starting material? Explain your answer.

_____

_____

_____

_____

3. Write the nuclear equations for the following reactions:

a. Carbon-12 combines with hydrogen-1 to form nitrogen-13.

_____

b. Curium-246 combines with carbon-12 to form nobelium-254 and four neutrons.

_____

c. Hydrogen-2 combines with hydrogen-3 to form helium-4 and a neutron.

_____

4. Write the complete nuclear equations for the following reactions:

a. $^{91}_{42}Mo$ undergoes positron emission.

_____

b. $^{6}_{2}He$ undergoes beta decay.

_____

c. $^{194}_{84}Po$ undergoes alpha decay.

_____

Name _____ Date _____ Class _____

    d. $^{129}_{55}$Cs undergoes electron capture.

_____

PROBLEMS   Write the answer on the line to the left. Show all your work in the space provided.

5. _____ It was shown in Section 22-1 of the text that a mass defect of 0.030 38 amu corresponds to a binding energy of 4.54×10^{-12} J. What binding energy would a mass defect of 0.015 amu yield?

6. _____ Iodine-131 has a half-life of 8.0 days; it is used in medical treatments for thyroid conditions. Determine how many days must elapse for a 0.80 mg sample of iodine-131 in the thyroid to decay to 0.10 mg.

7. Following is an incomplete nuclear fission equation:

$$^{235}_{92}U + ^{1}_{0}n \rightarrow ^{90}_{38}Sr + ^{141}_{54}Xe + x\,^{1}_{0}n + \text{energy}$$

_____ **a.** Determine the value of x in the above equation.

_____ **b.** The strontium-90 produced in the above reaction has a half-life of 28 years. What fraction of strontium-90 still remains in the environment 84 years after it is produced in the reactor?